www. sirnewtonsfruad. com

ISBN-13: 978-1535150965
ISBN-10: 1535150963

Author Peet (P.S.J.) Schutte

www sirnewtonsfraud .com

There is a good reason why this book is on offer as an e-book and is therefore limited by all the constraints accompanying such format. It is because I have an urgent message which I wish to share with students but I am handicapped in trying to do so. I am handicapped by publishers not willing to go against Newton and therefore go against the science establishment because if they do go against Newton, they have to go against the grain of the entire physics world. That left me one option, which I now am per suiting. I am giving this book for free in the e-book format because I do need to reach students everywhere and I do need to reach students urgently. However, the e-format also handicaps my ability to tell the message in its full entirety as the message should be told. This means the message is also handicapped in the e-book format and with that comes all the limitation that goes with this method of publishing that is also restraining some printing possibilities when using the e-book printing format. I have to use the option I now am following due to my personal position money-wise. I follow this procedure because of my personal lack of recourses in funding this project when using a normal publishing effort of application in printing when using paper and ink. The method of paper and ink is costly and Academics in influential circles are boycotting my work for reasons of protecting their personal work as well as protecting Mainstream Physics policy. In preventing me from telling my message they stone-wall any effort I try to achieve on my part to get work printed through normal channels. Due to my personal circumstances of fighting the powerful academics and also due to this fight I have with the Academic physics principles, I am therefore not forming part of Mainstream Physics. Since I fight a lonely battle I have a much limited budget and can't have the books published on my part in the usual manner. The financial restrain I suffer is largely because I have spent the last eight years researching science whereby I had to forfeit any regular income when I did the research on the work I offer as newly developed information. Whenever I reach out to have my work published, I run into an impenetrable wall that I can't break through.

Now that I have the answers and from which I have drawn the conclusions, I now find so much resistance from mainstream science in getting the findings my research uncovers out in the open. I offer for sale many other books in which I use diagrams, sketches, mathematical explanations and cosmic photos including other tools I employ to promote the required understanding needed to bring the ideas across that I wish to promote. However, publishing in this manner is very costly. As I said, money is one thing I do not have. My books carry this message of the fraud hidden by Mainstream Physics. The fraud is well hidden in physics. It is sand walled by those academics where they use the information they present as physics as if the facts they present are well proven principles and the manner of presenting these facts as such well proven facts became the teaching methods they use to fool the physics student about the truth.

What is the truth…when you have completed this book you will have had a peeping view, a tiny glimpse of the truth…but as little as you would gain from reading this book alone when put in comparison to what any person can gain from reading all of my work in total, you will gain endlessly more than what your tutors will allow you to see about the truth because what you have gained in reading this document is much more than what your teachers know about the truth. What I try to convey is that there is a good reason why academics block any and all publishing of my work, and when finishing this book, in comparison to what I offer, you have not even opened a first page of what I offer as new information when judging what my other work uncovers. Still, your effort in reading this document allows you to discover so much more of science's fraud than what previous students knew about being part of a brainwashing process in progress for generations. You will know so much more than what previous physics students had on offer.

The launching of this book has the purpose to show how much deceit there is going on in terms of the regular conduct of mainstream physics supplying normally accepted information in relation to what is proven and what is just fairy tale science. What you are taught as proven science is no more than fairy tales and that you will learn when you have completed this book. By informing students about these aspects I try to direct you into becoming a thinking student, and not a Newtonian stooge. I am getting you to ask questions about the fraud Mainstream Physics has been using for so many centuries. Unfortunately the new work my studies uncovered is of such a nature that it would not be possible to publish the truly informing work in this format (e-books), but as you are about to see, there is much ground for a lot of concern to have by the students studying physics.

Newton at first, centuries ago and then later also his followers committed blatant fraud. About that there is no question. However, in my quest to draw attention to the flaws that I detected, I managed to get no-where and it took me eight years to realise that the mistake I tried to uncover, was known all along by all the academics tutored in mainstream physics, but also was being protected by the same persons who was wittingly performing a cover up bigger than anything that the world ever saw before. This now becomes my attempt to bring the truth to the attention of students and help students to realise what the extent is of the fraud being committed as well as what fraud there is being used as everyday information and what is furthermore committed in order to disguise Newton's total lack of insight into physics. About this claim I challenge everyone who thinks Newton knew physics, to bring evidence that will prove me wrong.

Please keep in mind while judging my effort that in a book this size, I can bring very little of the overall argumentative proof required to underline the whole issue required to sanitize science. However I have written round about twenty plus unpublished works and when reading for example the four volumes of a book such as **An Open Letter On Gravity Parts 1 + 2 Volumes 1 + 2,** then only after that will such an effort ensure any reader to gain an understanding about the foundation of my work. Please note that I say when reading **An Open Letter On Gravity Parts 1 + 2 Volumes 1 + 2** it will bring **understanding about the basics** and only bring understanding about **the basics** of what my work entails. This which you are about to read is only 2 parts of the eighteen part frame on which I worked to develop **An Open Letter On Gravity Parts 1 + 2 Volumes 1 + 2** and this book must please be seen as such, it must be seen as two parts of eighteen I used as a framework in planning when I developed **An Open Letter On Gravity Parts 1 + 2 Volumes 1 + 2**.

The effort that this book represents in informing about an entire new way of cosmic appreciation is only but to show that there are grounds for concern and this book does not even bring all such arguments indicating concern in full. It aims to caution students about the motives academics have in brainwashing students to accept Newton. This effort aims only to warn students to look out because there are much more phoney science. My distress on this matter is so great that I am prepared to divulge this information for free as to warn students about the misconduct they are ruthlessly exposed to. However, telling about the new science as I uncover the fraud takes many volumes of writing since there are many volumes of fraud that has to be corrected. That one can only find when reading the first ten letters forming books named as with a title beginning with **Open Letters...**and those titles are included as books which I mention in this book

All books on offer for sale are subject to changes of a variety of reasons where reasons sometimes are not to be disclosed at every opportunity.

As I have pointed out, I have written a list of books but I could not find a publisher willing to take on the might of the Physics Institution as I have to do in the manner that I have to do when confronting the Newtonian institution... which is head on! Publishers use the advice of trusted academics that is well accepted by the rest of Mainstream Physics to advise the Publishing house on matters such as accuracy, acceptability and most of all, the commercial aspect such a book will present. This is all covered by the title as being the feasibility of the book's future, and there is not one academic that will ever advise a publisher to bring to print any work that goes against Newton. Because my work goes against Newton, my work stands no chance of being positively advised to be published. By going against Newton, I kill my book before its birth. You might see it as a publishing abortion just after conception. Going against Newton is going against the grain of physics, or so Mainstream Physics brainwash all to believe. Any academic advising a Publisher that my work must be reviewed positively will with that also destroy all other work ever published on physics, which would include the personal work of the expert person in physics doing the advising on my work. There just is no academic going to disclaim his or her personal work by positively crediting my work. However, at first I was so naïve that I thought academics were honest enough to publish the truth, whatever the truth would be...and was I wrong about the honest conduct of academics! I thought academics had a life long urge of discovering the truth about physics. As shocking as it was, it took me eight years to see that Physics makes a mockery of the truth and kills off anything that goes against Newton. Now I present you with the first work ever published (as far as I know) that not only goes against Newton but shows that Newton is a mockery of true physics. In doing that and by going against Newton killed my work instantly and every time. I had one option and that was to publish my work privately.

To go and privately publish a book such as those I refer to as being much more informative than the book you are reading, is very, very expensive. Those books I write about physics do not use your standard Mills and Boon Publishing printing layout because I use a lot of schematic sketches, drawings of all sorts and other graphics to explain the point/s I wish to make, and everything published extra going beyond the normal in the book is numbered in dollars ($) or pounds (£). Since I also do not have the financial recourses to have the other books published privately, I have written a long list of books waiting to be published at this point, for which there is no money. I decided to have this book published, in the e-book format, because after all, it is the students I wish to address, because if I could get the students aware of the corruption that is taking place forming part of their education, and I could get the students to bombard those academics with questions that will force the academics to come out of their dark hide-outs while they expose their crimes to the students, those persons ignoring me, will be unable to victimise the students. If those academics holding teaching positions could no longer victimise the students, because the students got wise to their fraud and started asking the correct questions that indicated the students finally became aware about the corruption, then those that do not wish to listen to me will then be forced, by the students to listen to what I say.

That (I dearly hope) would lead to persons getting more inquisitive about what the truth holds and I hope to sell personally-published work to those being inquisitive. With the income I hope to derive in this manner, I plan to invest to privately publish the more elaborate and much more informing books that show what remedy science has to undergo and what corrections Newton has to endure to stand corrected. The main issue is getting the message out to any and all that is willing to listen and moreover to those brave enough to confront Mainstream Physics and force them to prove what they maintain was proved by Sir Isaac Newton. Let them for the first time in the history of physics, go on and prove Sir Isaac Newton correct! By doing that while earning money, I then hope to gain a means of liquidity whereby I could have my truly informing books published commercially on the Print-On-Demand format and have those publications entered into the market through the normal commercial channels. I hope to direct such funding obtained from the selling of these books entitled **Isaac Newton: A Conspiracy to Defraud Science** as well as any others on the rest of the list as to then have the rest of the titles on the list published. However, there already are other books I mention which is already available at this point. Any of the other books mentioned in the part **www newtonsfraud http//users @lantic.net** can be obtained through my web site at the prices listed in this publication and in **www newtonsfraud http//users @lantic.net** and when purchasing the copies, the printing is done by me in person. The income of those books would be used to commercially print the others I earmarked to be commercially published. Also any profits gained from such selling will go to publishing the other books on the list. Should you wish to purchase any of the books privately, then please read the final part of the book that you are reading, or visit the web site **www newtonsfraud.com** that is devoted for such a purpose, but only do so when this book you received free of charge convinces you there is a lot of rotting cadavers hiding under the mantel of accepted mainstream physics. **If this book does not convince you about the reality that the work Newton introduced and the work that now forms the basics of physics has never been proven, then better be sure that you seek good Psychiatric help because you are a serious case of one that is already being highly brainwashed!**

Go on and ask your tutor when will the Moon finally hit the Earth as it has to if Newton's gravity of attraction by the mass thereof was ever correct! Remember that the gravity Newton introduced claims that the Moon and Earth is in gravity by attracting each other. Therefore, the two has to collide at some point, that is if Newton's law of attraction is true. This part of the attracting qualities of gravity no Newtonian ever suggests! Those I accuse of being Masters in physics and therefore forming a part of the cover-up of Newtonian fraud might even admit that the Moon will never collide with the Earth and planets will never collide with the Sun, but they will tell you that there is an ongoing study on this matter called Critical Density Theory. Well, if you think Newton was ever capable of fraud, then wait till I dissect this scam and show what true irregularities are in place being implemented to cover Newton's misconduct. You can't believe how blatant a person can be when those thinking they are more clever than all other humans, go about committing fraud by using Newton's laws as a smoke screen.

The purpose of this book and my motive in publishing this book is not to convince anyone about anything. To do that requires much information and room to do that is therefore not enough in a print such as this. The intention I have with this book is to show there are

irregularities and to address students to become aware of the misguided trust they place in academics. Those entrusted with the youth, should at least be honest people. This book has one intention, and that is to make students aware that there is reasonable doubt in matters academics presented as facts. Students should feel very concerned about such irregularities.

To be much better informed, then read Newton's Fraud, and also Sir Isaac Newton's: a Conspiracy To Defraud Science as well as the book in parts named as An Open Letter On Gravity Part 1 Volume 1+1 and Part 2 Volume 1+2. What you are about to read comprises of extractions forming two parts of the framework from which the actual book that holds the title as An Open Letter On Gravity Part 1 Volume 1+1 and Part 2 Volume 1+ was developed from. Use this information divulged in this book, as limited as it is, to test the reliability of your tutors' teachings. Confront them with facts and don't allow them to stupefy you with their ability to commit mind control by deceit. Remember, deceiving you are what they really are the Masters of and they will go on to blindfold you with more untruths in order to escape your uncovering their deceit. Insist on answers that bring results and do not allow them the opportunity only divert the attention away from their deception. They are not knowledgeable in giving answers about the correctness of science because they are not Masters in science **but they are the true Masters in deception**. When you confront them, **do not trust their honesty** as you have done all along, because if and when you do trust them, your trust will open the door for them not **to come clean with truth**, but to even blatantly bullshit you more, because that is what they are good at. If you doubt their sincerity or they try and convince you about their honesty, then ask them to show when the Moon will collide with the Earth and insist they use Newton's gravitational formula $F = G \dfrac{M_1 M_2}{r^2}$ **to prove that Newton is correct.** If they are unable to show what precise rate of attraction there is, and when the looming collision is going to take place between the Earth and the Moon, then again realise they bullshit you and **they convey distortion as truth…do not trust them**!

As I have mentioned, there are also more informative books available for the more inquisitive readers.

I am supplying a list of books waiting to be published as to show what the extent is that my research reveals as well as the extent of the rejection academics performed to ignore my work. In all the one sided correspondence I had with academics to this day, there still has to be one academic who was able to show the incorrectness there is in any degree on my part, how small that incorrectness might seem of any point I have raised my concerns about. These books I name at the beginning as well as the end and I do so to show that there was a huge study done on my part as to indicate to academics the totality of the problem about Newtonian science seen in the context of the overall picture.

To whom it may concern and all others reading this document:

I put this challenge to everyone entrusted in upholding Newtonian physics as well as those teaching Newton's physics at all the various levels that they may do:
PROVE ME TO BE INCORRECT WHEN I SAY THAT NEWTON IS EQUAL TO FRAUD!

Prove that the cosmos is growing smaller when using Newton's concept of gravity by attraction. In other words, with using available data that was recently obtained, prove where, when and how does the cosmos contract as is taught by Mainstream Physics when using

$$F = G \frac{M_1 M_2}{r^2}$$

and where, when and how does this formula really apply!

The distance between the Earth and the Moon was constantly measured to a micron on a daily basis for the past forty years. Let your tutors tell you how much did the Moon and the Earth move closer…in other words how true is Newton's formula applying in reality when precisely measured. **The Earth and Moon distance is expanding in the same manner as the entire cosmos expands and in the same way as the cosmos is doing! This is a well know but also well hidden fact no one in Mainstream Science ever openly discusses or divulge in the classes when teaching about Newtonian physics that sustains the principle of gravity being a force of contraction and using a value of mass.**

This is my introduction and this is my prologue:

But before I can commence with that task I have another duty to administer:

I do find much pride in my status as being Afrikaner and would like to have my names used by pronouncing it in the manner Afrikaans dictates…therefore I would sincerely appreciate the courtesy when readers will take note that my name and last name are pronounced in Afrikaans, which is originally from Dutch and must be pronounced in that way. Peet one would pronounce "here" which is the closest English to the pronouncing of the "ee". The "Sch" in Schutte is pronounced exactly as one would pronounce sch in school and the pronouncing dictates that it is done in such a manner as where both actually are pronounced Skutte or "skool". By pronouncing my name in Afrikaans you do me the utmost courtesy any one can. Being an Afrikaner is what I am most proud of.

If you are a student in physics then you should read the following information. One could think of another name for physics and that would be Newton's mythology. It is about the subject of gravity and is most important. The "Newton's Mythology" comes from the fact that students have to learn what never was proven. Do you realise that it is an accepted practise that all students who are studying physics on all levels are subjected to the most intense brainwashing and thought control found any where on Earth? This must be some sort of a joke you may think but thinking that way in disbelief is just what those practising the mind control wish you to think!

I came upon a mistake concerning physics.

This mistake is about the cosmic phenomena called gravity. Detecting the mistake is simple because it is uncomplicated to understand. Academics in Science say that a feather will fall with the same speed as what a large rock would fall. However in the formula of gravity $F = G \dfrac{M_1 M_2}{r^2}$, which is the formula by which all things fall, the driving force falls on the mass factor.

The feather falling as fast as the hammer or large rock is according to Galileo and that is accepted as a principle in physics. But Newtonians put this motto equal to the one Newton has about mass driving the fall. How can mass that puts differentiation between objects have objects of all sizes and forms fall equal? For the first time ever since the time Newton introduced gravity, I seem to be the first person who questions this interpretation.

Has anyone ever explained how the idea of how a feather can fall as fast as a hammer, fits into the idea that mass pulls mass by the gravity it initiates and how the falling by gravity forms power that is exerted by mass. This totally contradicts Galileo's concept of falling in equilibrium, as a hammer has much more mass than even a large feather has. If you are bedazzled by the feather idea because the feather shows wind resistance and rather float than it falls to the ground, then replace the feather with a truck and you will still get the truck falling equal to the hammer. The feather concept Newtonians use as a part of their fraud and I get into that in other more informing books. How on Earth do these two concepts of a feather falling equal to a hammer proved by Galileo, fit into this interpretation they use of mass causing objects to fall?

How can a large mass pull as equal as a small mass pulls to travel equal at the same speed over the same distance and still be driven by the power of mass creating gravity. Have you given this idea a good thought? By me scrutinising this concept I disagree and by me disagreeing I am silenced by those in power. When any person disagrees with any academic in any lecture hall about mass not forming a picture as being the factor responsible for pulling gravity and you come to a conclusion that you doubt the mass part that they bring into the picture as being responsible for establishing gravity, the academics wipe you from the table with a swipe because then they contemplate that you are so stupid you fail to see facts and you are too stupid to understand physics. They even in some cases go on to say physics is not for stupid people! They will throw you out of the class if you persist and insisting on an explanation, or they will fail you in some future examination and get you excluded from campus. This is part of the mind control. Either you allow them to force feed you inconsistencies and play along as they brainwash you by accepting their inconceivable irregularities they teach, or they get rid of you on the grounds that you are too incompetent to understand physics!

I have been at odds with academics for years and only because of the superior positions they hold in office are they able to bully me into silence. Academically I am not from their league and with me not being from their ranks and with me not being part of the mental sphere they claim they have; they are of the opinion that with me not forming any part in Mainstream Physics that then disqualifies me to have any opinion on the subject of physics. Being what they are gives them the right that they may regard or disregard all opinions when they do not fancy the opinion. They may silence whatever I may say notwithstanding my correctness and validity just because they are in an academic office. Absolute power corrupts absolutely and they are the living example of that.

Due to the important positions Academics hold in the huge academic institutions, such castles of power give them free sanctuary from where they can hide their criminal ploy of deceit. They do not need to explain anything but only hold accountability to themselves amongst themselves and their deeds go totally unchecked. That makes them to be the untouchable and unapproachable powerful from where they rule with absolute authority. This unquestionable authority gives them the locations, erects a cover and gives them the opportunity to hide behind that wall of absolute superiority from where they can suppress little persons such as I into silence and submission notwithstanding...Whatever I have to say can never go past their scrutiny and can never pass their sanctions.

What Newton saw as gravity can't withstand even the slightest test of proof and I showed that it is not possible to use Newton's formula as Newton suggested it applies to mathematically calculate gravity. I come back to this issue later on. I have tested Newton's thinking and the book I offer to you for investigation serves as the testimony to all the testing I did on Newton. This any body who can see and also will see when reading this book. I tested Newton from all the angles to see if he possibly could be correct but found his thinking wanting every time. The truth about Sir Isaac Newton's concepts I came to conclude, was that the reality is that it is not in any way overstated to declare that Newton conspired to defraud science and moreover that he committed blatant mathematical corruption in trying to prove the concept he had about what he thought forms gravity. There is no backing for Newton's ideas and even the ideas which are in use are not in the form that Newton said it applies where physics in daily use serves as the best discredit to Newton bringing no proof about any of the claims that Newton made on matters concerning science in cosmic gravity.

I show that every thought Newton introduced that later proved useful and was correct, was what he stole from another far better cosmologist called Johannes Kepler. Not one of his laws are directly relating to any concept Newton ever introduced at any stage but is the result of academic theft he committed against a much larger figure that preceded him by almost a century. But he stole, he lied and he raped the work of a predecessor in order to defraud the world of science in his time. Newton brought no original input into science except that he gave a concept the name "gravity" and even that is inappropriate. Newton made suggestions that break every mathematical principle he could think of. That, Newton did in his attempt to win over the prevailing academic thinking of the day in his time as to lay some sort of groundwork to form backing for his ideas on physics and to attempt to explain gravity or what he thought gravity is. If this is shocking and sounds outrageous, then a lot more shocking detail awaits the reader in this book.

Newton's claims about the principles he declared as being responsible for guiding physics carry no proof and after I realised that, I was able to start forming another line of thought on gravity. After formulating my concept about how gravity was truly formed, I had to introduce my ideas to academics in physics. In my quest to find the method how gravity formed I used the four phenomena and the principles of these phenomena as well as determining in which way each phenomenon applied. Not surprising is the fact that the shear existence and acceptance of the four cosmic phenomena is very securely hidden from normal investigation. Since Newtonians can't use Newton to explain the **Titius Bode law**, the **Roche limit**, the **Lagrangian positions** and the **Coanda Effect**, these phenomena are hidden so deep that it is very likely that you as the reader might never even have heard of these phenomena. It is because these phenomena portray Newton in a dim light and that no Newtonian would allow to happen! I use these phenomena to mathematically prove what gravity is, but the condition I apply is that gravity is not even close to what Newton suggested forms gravity. By my ability to prove gravity I had to place each one of the four in the way forming one part of what gravity is. Before I could manage that, I had to find a way as to explain what the working relation of the four cosmic regulating phenomena were and I had to figure out what was the known contributing facts to the cosmos that each was responsible in contributing to gravity in my effort in determining how they work and then implicated that specific formula's function mathematically in forming gravity in the cosmos. This was no easy task but I did it and by formula shows that my argument is logic and the mathematics prove that it works well.

The Official Policy Protectors of physics never try to explain the relation between Newton's laws as mentioned above, and the binary star system forming the principle we know as the Roche limit. According to Newton's formula, stars have to collide, but the Roche formula explicitly shows there can be no collisions between stars. This fact Newtonians hide with much deception. The binary stars are systems where two stars spin around each other and never collide, whereas taking notice of Newton's formula; they must run into each other with ever increasing speed. These stars are many times bigger than the size of our Sun and yet they fail to draw each other into a collision. If Newton's

law on gravity $F = G \dfrac{M_1 M_2}{r^2}$ did apply as Newton said it does, then the attraction between the

two by the mass of both had to present a serious and all demolishing collision. When one applies the same Newtonian formula as given above, these massive giants must crash into each other, destroying themselves in the process. The enormous mystery is not in the apparent misbehaviour of these giants, but the fact that this is known to science since the previous century. Then we have the comet as an example proving that quite the opposite of Newton's gravitational law

applies $F = G \dfrac{M_1 M_2}{r^2}$ exemplifies attraction applying most of all by the most massive and

least of all by the least massive. What I try to say is that in the case of the comet, the comet has the least mass but is also drawn around the Sun with having the biggest variation circle whereas the bigger planets hold orbit rock steady. If mass was responsible for gravity, then Jupiter should vary the most in orbit while it does not! Size is what should count and yet it is the comet which is the smallest and the least massive, is the object that is pulled the strongest while Jupiter being the biggest and most massive, orbits like clockwork and never moves towards the Sun.

With the comet, the Newtonians regard a force to attach to the sun in some way where this force pulls the comet towards the sun. At the same time another force joins in that pulls the sun closer to the comet, but such is the mass difference between the sun and the comet, the force the comet applies never realizes. In view of this, it then is only the force the Sun that applies, as it is only the comet coming closer to the Sun that comes into effect. The comet proves this force by speeding up its movement as it comes closer to the sun. If the force did not become greater, by the effect of the mass increasing since the dividing of the distance reduces, the comet would gain momentum and in that there is no proof!

With the arrival of the comet in the sun's domain, the Newtonians leave the argument to be. The mass on both ends did the pulling and no further examination is required as to how the story unfolds further. If by chance Newton was correct, then the sun applying the force should remain applying the force and the force should increase all the time, accelerating the comet to the point of splash down. We must all argue that gravity is a force, which pulls an object to the centre of the larger object wherever that centre may be. The very same force that pulls the Chinese down is pulling the Americans, and if not for the surface of the earth's intervention, the next world war would be between the Chinese and Americans for King and country, honour and glory and to find who has the most powerful gravity force that will provide space to live in. If it is not for the earth being in place and therefore stopping matter falling right through the earth because of conflicting forces on both sides of the earth, the Chinese and Americans will then have to establish border checkpoints in the centre of the earth. The checkpoints will indicate where the Chinese gravity meets the American gravity and by allowing the force of gravity to find borders, we will finally have world peace. The only problem is to find the position where the Chinese gravity meets the American gravity and the two forces nullify each other. Just think if the forces of gravity, and not man, will intervene to set border standards: that must be the answer we were always finding a question for. This is a study far too complex to bother the United Nations, so we can find a more suitable group to investigate this fact to bring about world peace.

I am personally part of Africa, born and bred in Africa as an Afrikaner. I know the African solution to such a problem. In Africa, those in power will appoint a committee to investigate and then wait for everyone to forget about the problem in investigation, as long as the Western Powers pay the committee to commit to a non-existing problem. Therefore, such a problem is far better solved in Africa, because those in any African government aim to receive maximum western aid but never aim to solve problems. The African aim is not to solve problems but to wait for time to make it go away by postponing the solution to become part of the unsolvable. The African way is to ask the west for aid in order to create another useless committee to become over paid and under worked, quite capable of dealing with any non-existing issue of any magnitude as long as the investigation will never find an answer. Then sit around at leisure for as long as it takes to have everyone's memory fail on details of the matter and wait each month for pay day to come for many years while the west is paying the committee to be bored until their pension dates arrive. This is not funny...it is the way things are done. After many years lapsed and no solution came to the table, by then no one would know the name of the committee and much less remember the problem the committed committee investigated. You might think I am joking but this is just recoil from what is coming from your Universities in dealing with physics problems. Remember the Newtonians are the Brainy Bunch, the best intellectual minds humans have on offer and those who could bust a dam wall with shear intellectual power. On the other hand, the Newtonians are doing quite fine by their method on their own using a technique they apply for three and a half centuries. To solve such problem as physics are encountering with a Universe that is not respecting Newton for his shear accuracy, the Newtonians will apply a very different solution: Blame gravity's boundaries on a non-existing force, brainwash all future students in accepting it to be a force by telling them they will accept the force and forget the problem or fail the examinations and be chucked from campus, because that solved the problem so far. By the time, the student reaches a senior position he (or she) will no longer bother their mighty brainpower with the little aspects such as comets not falling into the Sun through the attracting force of gravity. They will advance to a point where they can mathematically move Black Holes around by creating Space worms, travel at the speed of light from galactica to galactica, and divert time back to the past while others calculate all the mass seen and unseen in the Universe and any other ridiculous notion they may find to test their personal brilliance. Those are the big issues that require the thinking abilities of the Brainy Bunch and not such mediocre thinking as to why the comet fails to hit the Sun. If you for one second think everything about this last paragraph was silly, the silliness started with thinking of GRAVITY ON BOTH SIDES OF THE WORLD, opposing each other, and that idea is not mine!

This is where the century old trick of the Newtonians work best; in case of finding Newton not substantiated by any lack of proof, then do not think any further and no further problem will arise. Leave gravity to be an attracting force because with a force and thoughtlessness applying even-handedly, the problem never surfaced yet and that continued for the past four hundred years or so. So why bother with a problem that bothers no one! When a fellow like Hubble proves quite the opposite to what Newton claim to be taking place with attraction by mass causing gravity, get a man who has a bigger ego than a brain such as Albert Einstein had and tell him to measure all the mass of every object contained by the Universe. Don't even think of how ridiculous the task is or how pathetically impossible it is for any person to even think of attempting to measure the entire mass offered in the Universe. Doing this, deception will keep everyone involved and occupied without doing something senseless while the problem vanishes through the many centuries to come. Ignore all the signs the Universe offer, to show how the cosmos defy Newton and remain resilient about gravity being a force of attraction, and mostly hold dear to your heart the realisation that the way of all forces are mysterious, but then also never admit in believing in magical powers and when doing all of that, you will be a great Newtonian my son. Those who do not accept forces to be of a mysterious nature should just contact astrologists and come to their senses about forces being not understandable. With everyone in agreement about forces, their nature and unpredictability, who then needs more real problems to solve?

No big-brain should bother about little issues like comets when there are so many galactica to conquer. But in my case apparently the comet-problem just would not disappear. I couldn't be a good Newtonian because I was handicapped by a small issue such as what we experience with the behaviour of the comet. The comet does not hit the Sun but speeds off into the dark yonder. In doing that (breaking loose and speeding into the dark yonder) something broke the force, something interrupted gravity. Let us see what happens. A force means it acts the same way as tying a rope on one object and start hauling the tied object in. The longer the rope is, the less control will be on the lesser star. As the rope shortens, the better the control will become. By implying that gravity is the force, our Newtonians tell us that we have to regard gravity in the same way the rope is hauling in a comet. It is something like fishing where the comet pulls, and the sun pulls and eventually the angler gets his fish. One may argue that the rope is not the force because the force is actually the hauling, or shortening of the rope. I have had Newtonians trying to avert the problem they refuse to see by bringing in this argument. This manner of reasoning has the same value as introducing the African committee of investigation that will never uncover an answer. The rope is the extension of the force in a way being the sole representative of the force and the instigation acting out the force. The rope therefore is the force, extended somewhat, but still acting out the application of the force.

Whether the rope eventually broke, or the hauling stopped, the effect as far as gravity applying its force, the process came to discontinue … and we know that gravity is a force that pulls something to the centre of the body in control of the gravity. What made the force act in defiance of its nature? Why did gravity change its mind? What stopped the sun applying its veracious onslaught of the body holding the poor defenceless comet? Non – Newtonians will blame me for exaggerating, but I know that there is no Newtonian that can understand my argument. In that light, I ask non-Newtonians to show patience, because there may be a few Newtonians that will also read the book and to them everything said this far does not make sense.

This is where tutoring comes in best. Should a student bring up such an un-academic and spiteful thinking about the mysteries of a force, then the lecturer sets a date on testing all the students' reaction about how much they accept the force. When any student shows signs of defying the force, the lecturer can fail him outright and have a good reason to drive the silly youngster from campus.

By ignoring the problem as to why this comet breaks free from the gravity of the sun, and continue in its freedom until gravity is at a point where it is most weak, may not bring answers, but it surely avoids nutty questions! Questions are not there to interrupt Newton's laws! Ask any Newtonian High Priest and he will either tell you that in a very roundabout way or he will simply ignore you by telling you to your face that you are incapable of understanding Newton. The best way to get out of the answer of course is to tell the sod with all the questions he has not the qualifications or the mental capacity to understand Newton. That will make the pest retract to some ditch he should be in, in the first place without bothering the greater minds with some stupid minor issue. How do I know this you may ask? I have been down that alley many times and have been treated with that precise treatment on occasions more than I care to remember.

Still, the comet defies the force of gravity and my questions remain unanswered.

Dear reader, if you wish to read the funnies, jokes and laughs -a- minute, treat yourself to some real good clean jokes. Read the Newtonians explanation about how comets came about; how they get to the sun and where they came from. It is going from the ridiculous to the thoughtless and ending in the realm of the mindless. However, be warned! Only do this on occasions where you feel very depressed. The jokes will otherwise drive you in a state of laughing hysteria. Poor old Newton was considered a very dry humourless chap in his day. To think what silly ideas can come from his forces.

Hauling in and releasing something caught on a line is called playing with fish. We might say that Newtonians love fishing and confuse planets, comets and fishes when they regard the interaction of comets with the sun. There is only one small problem with that argument and that is that fisherman and fishes form part of a second natural force named life. Life stands apart from the cosmos. Life and the cosmos only share time in space, not a joining of forces. Beside that, comets were part of the cosmos long before life had any role to play, so blaming it on some way life interacts with life does not cover the solution.

Why would the comet break free from the sun's gravity? That is defying the law of gravity. Far worse than that still, is the fact that the comet's actions have the nerve to defy Newton. No one alive can defy Newton and remain alive. Does the comet not realize its actions contradict the all- important Newton and the gospel of the Newtonian - Priesthood. The best way the Priesthood of Newtonian gospel can deal with such defiance is to ignore it and no one will notice the actions of the comet. That is the scientific approach. Ignore and forget the problem. It is as simple as that.

With that let us conclude comets and really enter the world of forces at work! Let us now apply our attention to the forces at work on the forces structuring planets. The next formula is very simple to understand. It is the fight of understanding the application of the formula that becomes not applied that is troublesome. If you are one of the few fortunate ones that are able to understand the applying

of the working of the force of gravity $F = G \dfrac{M_1 M_2}{r^2}$ formed by attraction of the mass and never

spotted it not working, wile ignoring the fact that planets have not moved closer to the Sun for the past five centuries, then forget the fact that you have been not noticing important issues and stay in the haze you enjoy that much. You then obviously are a brainwashed Newtonian and if not are a brainwashed Newtonian..., well I regret to inform you but there s a vital IQ factor shortfall. On the other hand if you only realise this and you feel cheated by those persons you trusted, then there is still hope that you have a clear mind left. The resentment you carry with you from childhood about the formula is in, not understanding it, but accepting the outcome of the formula you never could understand. Newton said that the force between two objects depends on the mass of both objects multiplied with each other and with the gravitational constant and the derived product you divide by the radial distance square that separates the objects. I shall put this in a mathematical language for your enjoyment that will explain the life-long "not understanding" to better effect.

$F = G \dfrac{M_1 M_2}{r^2}$. What does this say? The greatness of the force depends on the masses of

the two orbiting objects, aligning that product with the contribution of the gravitational constant. This then, you divide by the square of the distance between them at any given point. Please, in all fairness to you, the reader, I have to warn you that quite a number of professors in physics told me that by reasoning in the manner I do, I only prove that I know nothing about Newton and understood even less about his work. Considering such allegations, I shall explain to you what I understand in as much as telling you what I know.

The Newtonian formula states that the force between the planet and the sun will improve as the mass of the planet increases (becomes bigger) and by multiplying that with the universal gravity constant you will get a value that will become lesser, the larger the distance is between the sun and the revolving planet. With the reducing of the distance, the mass on either side must therefore be on the increase because it holds an inverted relevancy. This means the sun is pulling according to its mass. The planet is pulling according to its mass. The gravitational constant is influencing the pull evenly at both ends and the distance between the objects will reduce to the square value of the force's total application. I could never see what part I do not know and what I did not understand. No professor ever explained to me what it was that I did not understand either. That left me in a place where I did not understand what I did not understand and I never could see what I never could see. I shall try and make sense of my not understanding as follows:

This is like having two balls attached by a rope on a floor that holds the same drag on both balls. When I reduce the length of the string, the bigger ball will show a greater resistance than the smaller ball, therefore the larger ball will apply a larger tug than that of the smaller ball. The rope will reduce (become shorter) more at the end where the larger ball is than at the point of attachment where the smaller ball is. What is wrong with my argument? When the two balls are so miss- matched in mass as is the case with the sun and the comet, the one ball will do all the moving, leaving the larger ball stationary. Surely the tugging at the larger end must bring the smaller object closer. By comparing the mass differences, you will find there is no comparison. The smaller object just has to come closer with the application of such a force as gravity. We know that gravity can really pull. By standing on a tall building you will find proof of this. Drop a tennis ball down from the building's roof and see for yourself how it falls. The distance between the Earth and the ball reduces by some speed. With that being obvious, the distance between the sun and any planet has to reduce as the planet orbits the sun each year. Even if it is small, there has to be a visible reduction after four and a half billion years of pulling and tugging! Today after wrestling this problem for the duration of twenty-five years I can say (with a clear mind) I finally know how it works. It does not work!

In the beginning right at the start of my investigations I was always looking for mistakes on my part. At first, I thought there is a fifth force that I am unaware of because of my slender education, a force the academics can obviously see, but I cannot through obvious lack of education. I thought that my personal ill understanding of Newtonian the genius prevailing (something I was told that I suffer from) gave me a blind spot that every non-educated have and was born with. The blind spot cleared only thorough education as education removes it in the way only education in science brings knowledge. I thought the removing process similar to the way washing removes stains and spots from whites; education can remove blind spots through the process of intensive tutoring. All I wished for was some academic to help me remove my blind spot about comets and their behaviour. The comet's behaviour, I could see, was an exaggeration of orbiting patterns applied from our planets orbiting around the sun; in the way, we observe galactica in the sky.

Then finally I came to the point of accepting defeat. It was not I, with the blind spot; it was all the academics brainwashed into a state of having such a blind spot. Science insists on repeatedly ignoring mathematical principles, because Newton had his claim to fame with one single calculation, THAT HE, IN FACT, DISCARDED, BY THROWING IT AWAY. He made a brief calculation as a young man who saw an apple fall from a tree. Seeing this he jotted down a formula and then chucked it away. His piers and elders picked up the trashed paper with the calculation, and got all excited by

the logic implication it had. $F = \dfrac{r^2}{M_1 M_2}$. The mass of the two objects destroys the radius

between the objects. Everyone went ballistic, proclaiming him as an instant genius, the one the world was waiting for after the crucifixion event.

I do not, for one second, deny or dispute the revelation. What I do encourage is place the event into its correct context. It was merely, and simply an apple, that fell from its branch to its roots. The apple did not pretend to be a meteorite that fell from the heavens. If it were a meteorite, I am sure, with the man's genius, science would be somewhat different at this stage. However, as a young man, being very impressionable, as all young men are, and with the attention this brought about in the world of science, the matter overshadowed the fact. I am not disputing Newton; I am disputing the relevance of Newton's scientific breakthrough. It was not two objects of cosmic proportions, colliding in a show of spectacular. It was, after all, only an apple falling from a tree. With this miracle revealed, Newton found he was competent to improve on the work of Kepler and if I may dare say this, there must have been some political agenda behind this act and the accepting of it, for Kepler was a German and what German can ever teach any Brit! The very same politics are still the order of the day forming international rivalry on all fronts.

Newton, and science, made one enormous blunder, from this stance. They took the radius of a wheel not to have any influence on the wheel. In doing that, they removed the very fact that keeps the universal attachment together. They put two objects in a attaching relevancy and then announced no relevancy. Doing that is breaking the most fundamental mathematical principle.

$$\frac{dJ}{dt} = 0$$

This disputes mathematics. DJ / dt can have any number from eternity to infinity, only excluding one; it cannot be 0. By placing the one in division of the other, you bring in relevance. You cannot then say there is no relevance. By doing such, you proclaim that one of the factors is non-existent.

$$\frac{dJ}{0} = dt \text{ or } \frac{0}{dt} = dJ$$

In both cases, one of the factors then does not exist. Such a claim is incoherent, because you proclaim that a circle has no radius, or a radius has no circle. When calculating a circle, you multiply either the square of the radius by Π, or the quarter of the diameter at a square by Π.

$$\frac{dJ}{dt} = 0 \text{ constitutes a circle and is also therefore } \Pi \times r^2 = CIRCLE$$

If you remove r it then is $\Pi \times r^2 / r^2 = CIRCLE$.

You cannot then say $r^2/r^2 = 0$ and therefore $\Pi \times 0 = 0$. That is nonsense. $\Pi r^2/r^2$ will always be $\Pi \times 1$, and that is the eternal circle.

When looking at any rotating object, there has to be a point of no rotation and no rotation means "no rotation", not no existence. No rotation means a factor of 1, not zero. That then is singularity. The eternal Π, the Π that may not have significance but still it is a Π of value.

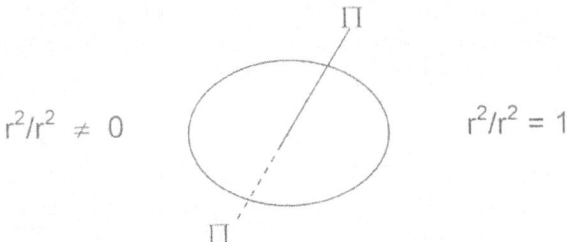

The relativity remains one, eternally one, but it cannot be zero. Therefore, dJ/dt cannot be zero.

dJ/dt can be eternal or infinitive or at the worst it can be dJ/dt =1 but dJ/dt ≠ 0

When explaining this to any child, they can immediately see that. Explain this to any Newtonian High Priest and he may have you removed forcefully from campus. I cannot find one Newtonian, large or small to accept that. By not having a wheel rotate, the rotation seizes, as it is not the wheel that vanishes into thin air. When the wheel begins to rotate, you cannot state that all things remained as it was. With the wheel in non-rotation the rotation still exists forming the infinite possibility of rotation. Then afterwards the wheel starts to rotate and by the start of rotation the circumstances surrounding the wheel changes. A wheel in rotation is very different from a wheel not rotating and therefore cannot be the same thing. By establishing non-rotation, the wheel becomes the factor of one, and the rotating action becomes zero. The wheel does not disappear. But in the same manner does a wheel in rotation not remain still. Moreover, the wheel did spin and that is something that all of Newton's cheating can't remove.

In the cosmos, everything is rotating because nothing ever stands still. Therefore the mean equilibrium, the common factor there is to share, has to be one, eternity, the eternal Π, because all rotating objects has Π in singularity, and sharing singularity, gives every object in space a relation with all other objects in space. After trying for many years to bring our Brainy Bunch the candle, I concluded that Newtonians are incapable of realizing that mathematical principle as a reality. They maintain they know mathematical principles far better than an ill literate such as I and yet ….

The comet rotates the sun, and the sun by itself has a point of singularity where Π remains without r. The comet, holding the orbit, also has a point of singularity, but since there is space separating the two objects, they cannot share a mean point of singularity, the very point of existing. Since singularity means just that, being single, there cannot be two. The comet and the sun have a mean point of singularity but the space they occupy divides their common singularity. That is why they orbit in an oval path, a path where the one structure holds on to more space from its point of singularity towards the space it claims. Since they do not claim equal space, BY THE DENSITY they hold, the space will not be in proportion.

They do share in the common fact of singularity and singularity cannot be two, because then it will be "dualarity" or (in case there is no such a word) duplicity where both find the space they occupy, with the space they hold, will be their individual eccentricity from singularity. The two objects are holding eccentric space around their individual but common singularity forming a point of mutual singularity in accordance with the individual singularity both claim space from. That point of singularity is Π the circle without the radius because the singularity removes all forms or values of r, leaving Π to be singularity.

That is why Newton is bullshit, and his $$F = G \frac{M_1 M_2}{r^2}$$ is utter nonsense. The moment you say Newton or any of Newton's laws, the Newtonian brain stun. For all the life in me, I could not once find one single Newtonian to see this. If you say Newton is wrong, they spiral down to become hysterical and just mention gravity and they all fall on their knees, cover their eyes in the ground, start praying and you cannot make them say anything other than hearing them mumbling repeatedly about Newton always being correct. Dare say there is no such a thing as gravity and Newton is wrong, they have you in an armed escort patrol, straight to the department of mental disabilities and psycho diseases in preventing you committing acts of extremely dangerous life threatening behaviour to yourself and others.

What is it the Newtonians fail to see? If an electron is orbiting around an atom, the inside of the atom must be a circle. If the atom was not a circle, it then had to be a cube. The electron cannot rotate around a cube; therefore, the inside of the atom is a circle.

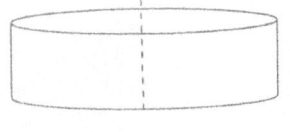 In a circle, there is a radius that initiates the circle. The calculation of such a circle is Π X r^2.

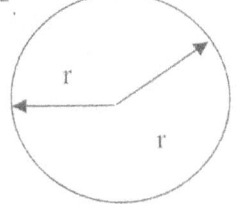

The radius r runs from the circle outwards, from a circle centre point towards Π, the value of the circle. In the centre of the circle, there is a point where the radius starts. It runs outwards from that point in all directions towards the circle Π. Technically, there then has to be a point where r is infinite and not zero, an absolute infinite. However, the circle therefore remains Π. The circle does not disappear; it remains there for all to see. It is only the radius that almost disappears into the infinite, but it does never become zero!

$$\frac{\Pi r^2}{r^2} = \Pi$$

If one removes the radius from the circle, the circle remains, only holding the value of Π. By removing the value of r, Π becomes singularity with no place to be. Singularity is the place where there is no space to be in place. However, Π remains because once r receives the slightest of space Π will find space. Then the circle will grow to Πr^2 and r would determine the space. Without space, there is no r but there is a circle with the value of Π. Singularity is in every single rotating object, be it the proton or the combining effort of all particles in the universe. That is what light and the photon is.

It is concentrated heat that the sun (or any other generator of electricity) connects heat to singularity where the heat receives either temporary connection to singularity or a small piece of individual singularity.

At first you as the reader may think I am trying to create a mountain from an ant heap, but in scientific terms the human race is preparing for the start of the cosmic journey. By completion of this book you will realize how "Xepted Science" (a term that I introduce to name the devious Newtonian science that only finds proof to further deception) believes it is built on a solid foundation, and, oh boy is everybody in for a rude awakening. Compared to the leaning tower of Pizza, science is about to start with the next section of a much bigger building adding many levels and already the view at the bottom where I am looks far worse than the leaning tower does.

If I contacted and argued with many South African Physics Lectors or Professor about Newton, I have been in correspondence with at least a couple of hundred. I must admit, the correspondence was overwhelming one-sided as the lot never once replied on my correspondence. What prompted my

initial inquisitiveness was the comet's orbit and reluctance to adhere to $F = G \dfrac{M_1 M_2}{r^2}$. If this

is correct to the last letter, then there should not be comets left for the lot should by now have landed in the Sun! The commit truly fascinated me from my childhood days, in the way it defies all the laws of gravity. Since my very young days, I was in search of what I at first believed to be a fifth force. I have raised the argument with just as many people not schooled in the art of physics and received a very different response. The most amazing aspect was the fact that the two groups were that far polarized. The non-physics group reacted astonished, amazed, disbelieving and reserved about my view about comets at first, but with their distrust not withstanding, everyone saw my point. The non-Educated responded in the same manner that I did at first. They argued that I was missing something of vital importance because "why do the wise not see it", was their argument. They always were of the opinion that I was too little educated to understand, while the educated was of the opinion that I was too little educated to understand. Neither party had the same view about my not understanding. The non-Educated understood my argument, but dismissed it on the fact that it was so obvious, I missed the rest of the knowledge behind the facts that made my arguments too difficult to understand while the educated dismissed my argument that I could not see anything which they could not see. Education brings the ability, which then made me unable.

 In short, they thought I was too stupid to know the rest of the story. Polarized to the non-Academic view was Official Policy Protectors where not one academic could understand my argument. The academic response was as much defending the Newtonian view as it was drawing a blank about my questions. They all seemed as if their ability in understanding my view was completely locked behind some wall. The non-Educated, of which I am a member, at least understood what I was saying, but dismissed the simplicity about the argument. In the corner of the Official Policy Protectors, was no response of any kind, but to feverishly defend Newton by raising the dumbest arguments I have heard. The arguments, even the most highly educated brought to the table, seemed motorized and non-responsive. When it seemed that their acceptance of the points I raise with my questions will demise their senses, then in defence, they put up a block! There is a peculiar sense of numbness in the way they could not understand what I did not understand. The academics showed no signs to indicate that they could even argue my point of view, by responding that I have an argument, and from that launched a responding argument to explain how or where I made my mistake. Their abilities in even understanding always seemed to hide behind a wall of not understanding that someone may not approve of Newton's arguments.

Newton says two pieces of rock will draw each other closer by reducing the distance keeping them apart. That we all can see by merely jumping in the air. No sooner have you lift off than you are back on the ground. That is what Newton said about three hundred and fifty years ago. Even trying to tell the Official Policy Protectors that Galileo said mass of an object has nothing to do with the falling, seemed to pass the Official Policy Protectors' sense of comprehension by miles. I was told on so many occasions that I did not understand Newton, but there it stopped. No one could explain to me what it was that I did not understand about the comet missing the sun by miles, where it was supposed to hit the sun with a dazzling impact. On this point, I cannot get through to them as much as they cannot get through to me. Our understanding is so far apart, we do not share the same planet, and yet after all my arguments and investigation no one and I repeat: not one could once clearly tell me what it is that I do not understand.

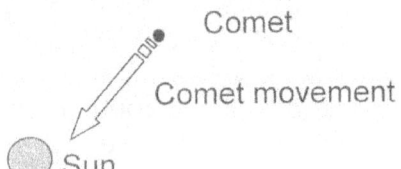

You have the sun and you have a tiny piece of rock covered by water also better known as the comet. There are thousands of them flying around, but never aimlessly. At first Newton's formula makes pretty much sense. The sun draws the comet towards the sun, as Newton said it does. The comet responds by speeding towards the sun, also as Newton predicted. Anyone can see a collision coming ten miles away. The sun applied gravity, the comet applied gravity, the sun is far too massive to fly to the comet, so the comet with much less mass does the flying on behalf of both objects. Every person with even the least of knowledge about science knows how the gravity application works.

The gravity of the sun collected the comet from no-one knows where, pulled it through billions of kilometres to the area where the sun produces the gravity with which it pulls the comet where the comet is to find its last resting place. The mass of the sun is obviously so large, it could produce gravity that can locate any comet hiding anywhere and collect it as a souvenir. What is there to understand?

The gravity of an object always points directly towards the centre of the object, the very, very middle point. Concluding from the fact that the comet is heading towards the centre of the sun, just as much as the sun is heading towards the centre of the comet, would not be out of line. The two centre points are heading for a direct collision, the collision becomes more and more unavoidable as the radius reduces by the value of the gravity that the mass of the object produces in accordance with the gravitation constant. The comet is heading towards the sun, and by not even moving, the sun is moving towards the comet, by attaching the movement the sun were supposed to have, on the comet. Newton's law proves to be exceptionally correct.

As the sun/comet, radius reduces, the radius separating the mass of the sun and comet effectively increases the relativity of the mass influence on each other in the form of gravity. The mass of the Sun and the comet increases by the factor of reduction of the radius separating the two objects. That will produce a growing gravity force as the comet / sun radius becomes smaller. By the time, the radius becomes one, the mass will grow on either side by a relevancy of 100, and when the radius becomes infinitely small, the relevance to the mass of both structures will raise a force with eternal power.

At a point, where the comet / sun apply a force of immeasurable strength, the comet breaks this immeasurable force. Remember the direction of gravity always points to the centre of the object, and that is where the collision is heading. As the objects draw closer, the distance reduces, but in accordance to the relevance the objects also become that much bigger in drawing power. It depends how one considers the relevancy to grow by the approaching comet which is by coming closer, diminishing the distance between the objects.

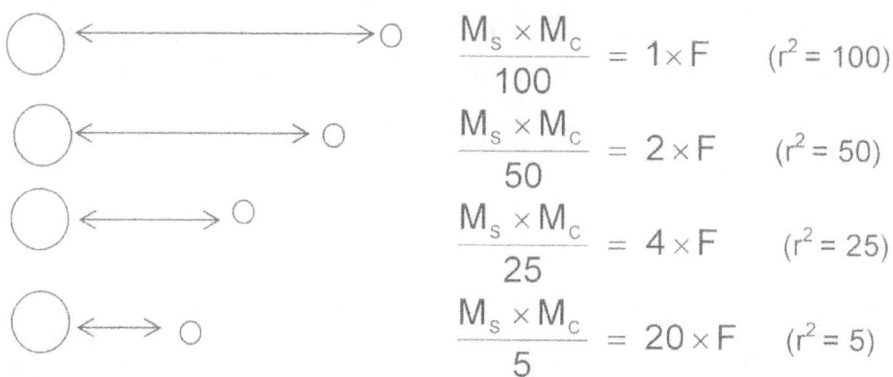

$$\frac{M_s \times M_c}{100} = 1 \times F \qquad (r^2 = 100)$$

$$\frac{M_s \times M_c}{50} = 2 \times F \qquad (r^2 = 50)$$

$$\frac{M_s \times M_c}{25} = 4 \times F \qquad (r^2 = 25)$$

$$\frac{M_s \times M_c}{5} = 20 \times F \qquad (r^2 = 5)$$

Then out of the blue, the comet finds the ability to eliminate the eternal powerful force of gravity, and keep at a safe distance around the sun. At this point, Newton goes sour. Nothing Newton predicted is happening. The comet and sun not only stabilized the force, the force begins to decrease as the radius between the comet and the sun is on the increase AT THE POINT WHERE THE FORCE IS THE STRONGEST, THE COMET BREAKS FREE AND SLIPS AROUND THE SUN, UNSCATHED.

Up to this point I can see the gravity having some pulling power with the comet coming close to the Sun. Therefore, I still see what is that the Brainy Bunch and other Newtonians see, but it is the way the comet's future progress from this point onwards that I fail to see any attraction playing much of a crucial part. From this point onwards the comet and $F = G\dfrac{M_1 M_2}{r^2}$ just don't match.

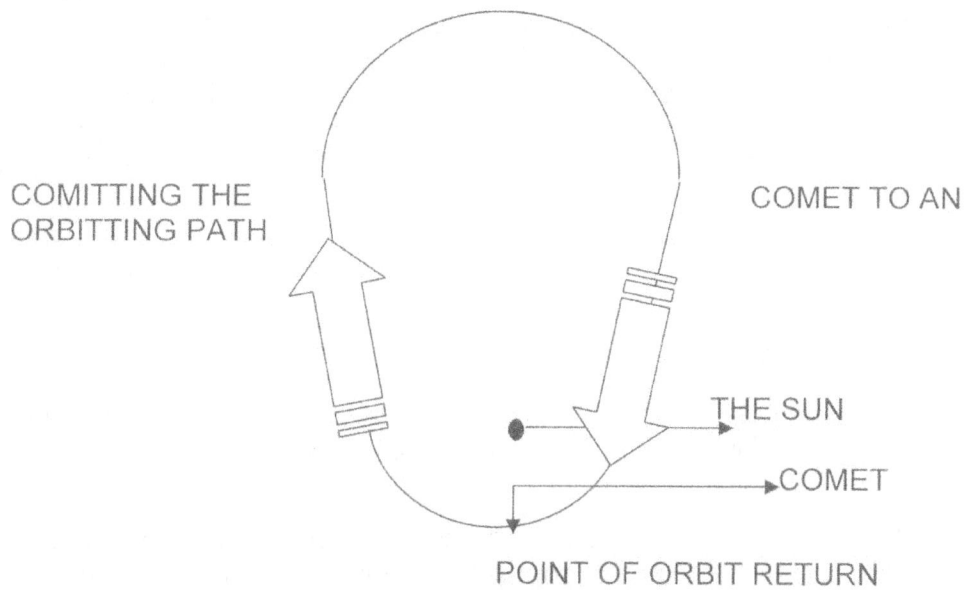

At the point where the force was the greatest, the comet overcame the force, but where the force was the weakest, the force overcame the comet's rebellion.

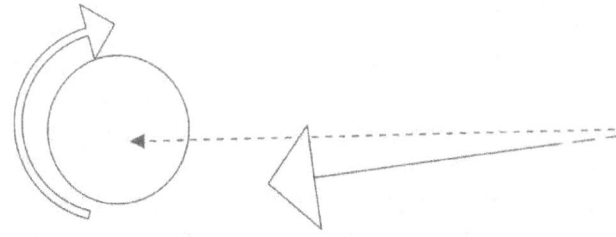

Then, in complete defiance of the Newton Law on gravity, quite the opposite applies. At the point where the radius that is separating the two cosmic objects is at its strongest, proves that the gravity force is at its weakest. At the point of almost no ability the gravity force suddenly releases enough strength to break, resulting in the parting of the two structures. The force now curbs the rebel comet on its way escaping the sun's gravity for the very last time.

The correctness of my argument is no longer the issue. The comet fails to hit the Sun and this trashes the correctness that $F \;=\; G\, \dfrac{M_1 M_2}{r^2}$ should offer. It was still a problem some twenty-five years ago when the Brainy Bunch could bully me into submissiveness…. It was at a time when I still held the impression that I was missing some point here. I do not state this phenomenon any longer in the hope of bringing across some flaw in my understanding. The flaw in my argument is not there because the flaw is science as a whole.

I could never understand the reason why "the ordinary", like me and others with my development level, can see what I can see, yet academics that have more brainpower in their heads, than I have life in my body, were unable to see such an obvious conclusion. You, those Official Policy Protectors are my superior in every sense a human can have, with the brainpower to break a wall, and yet you cannot see how far the tower of Pizza is leaning over.

I make the point to help you the reader to judge for yourself. If you are able to see the validity in my argument, you are not brain dead. Education has not yet bashed your thinking ability out of your scull. However, if a cloak of not understanding role over your brain, and numbness sets in on your ability to reason about this phenomenon, beware, you are a Newtonian. Newtonians should read this book very slowly because the effort you are about to launch, may be the most painful you shall ever experience throughout your academic career. You are going to suffer from reconditioning and Newtonian withdrawal, not that dissimilar to that of an addict in rehabilitation. You are going to reject me, hate me, despise me, loath me as you never felt about anybody else. If you think I am sarcastic, I am not. You will reach a point where you will abandon the reading of the book. You have my sincere sympathy and with all the soothing it may bring, know that you are not the first I saw getting such painful Newtonian rejection.

Once more, this phenomenon should not occur with Newton's presumptions about gravity. These bodies will and must collide and destruct, without a doubt. When the formula $F \;=\; \dfrac{M_1 M_2}{r^2}\, G$ apply, there should not be any force which is able to keep them apart especially when r reduces to almost infinity compared to what it is at maximum. However, they do exist and what is more, they maintain a certain distance apart.

With the "force" of "gravity" "pulling" the stars closer using the accumulative mass of the stars and multiplying that value with both objects by the mass component, this will reduce the radius r^2 progressively until r^2 reduces to zero. Seen from this view, it is little wonder that the significance of this was lost in the notion that this is yet another "mystery" of the universe. The scientists of the day (and the past) lost the importance, which this holds for us as earthly dwellers.

When raising these above mentioned issues, Newtonians will water down the seriousness. They will make as if it has laughably little worth. Be careful, because that is their brainwashing. That is the disinformation and that is the fraud they commit. They will say so what, everyone knows comets don't fall into the Sun and what is so surprising about it. What is so urgent about the fact that comets miss the Sun by a country mile to disappear into the black yonder. Claiming this as serious is shouting wolf in the centre of New York City and it shows that the person (I or me, whichever one you may choose) the one that is doing the shouting, is just a mad alarmist trying to steal thunder from a bursting balloon. They will say it has been obvious for centuries that comets miss the Sun and the comet missing the Sun don't say anything. That is their big scam! That is the serious corruption! That is part of the methodical brainwashing. This is part of what forms their fraudulent ploy. Minimising the error into non existence is what they are good at. Cheating the truth by fabricating a lie and diverting the truth is the scam they employ to get students brainwashed. They never prove that comets fall into the Sun or say by how much planets are closing in on the Sun.

The cosmos is expanding according to the findings of Hubble while they (the Newtonians) claim Newton is correct about gravity being a force of attraction. They use this finding of Hubble to date the age of the Universe. That is how accurate they assess Hubble's expanding or also called the Hubble constant is. But this contradicts Newton's attraction again completely. The Universe is expanding while Newton said gravity is attracting and then somewhere this does not match. But in the Newtonian mind they find a way how it will match. They simply commit more fraud and call it The Critical Density Theory. They will not even place the incorrectness of the attraction theory squarely on the shoulders of Newton, but place this dilemma in the quarters of the cosmos. It is the cosmos that is incorrect by expanding because it can't be Newton having the incorrect vision that the cosmos is attracting and therefore the cosmos has to carry the blame of contradicting Newton by expanding. Therefore this expanding must be portrayed as that the cosmos is simply going in the wrong direction for the time being! The cosmos will somewhere in the future, mend its incorrect ways and adhere to Newton's laws. Since Newton stands blameless and s always correct, the Universe must therefore carry the blame of this contradiction and therefore the cosmos better see to it that the Universe finds sufficient mass to start with Newton's attracting gravity. If it does not, the Newtonians will then be forced to invent dark matter that no one can see, in order to correct the cosmos and vindicate Newton! This blaming game and throwing the attention away from Newton's inaccurate assessment onto the cosmos shows how far Newtonians will go to cover up Newton's fraud. The in-discrepancy there is between Newton's incorrectness and the cosmos expanding will be forced onto the cosmos as if it is the cosmos that transgressed. Newtonians will rather divert of the incorrectness onto the cosmos and this scam they call The Critical Density Theory. In other books I show just how big a fraud this one is. They never prove Newton and when the Universe disproves Newton they minimise this as some joke a freak will listen to because it has no importance because all that anyone should care about is that Newton is correct and mass does the attraction.

It reality is either $F = G \dfrac{M_1 M_2}{r^2}$ applying or not applying. If Newton's surmising is applying and attraction gravity by the measure of mass is true then the comet should hit the Sun every time, or if $F = G \dfrac{M_1 M_2}{r^2}$ is a lie then this is proven by the fact that the comet misses the Sun every time the comet circles around the Sun. There can't be truth in the force of attraction expressed as $F = G \dfrac{M_1 M_2}{r^2}$ when according to the formula everything collides with anything where all that has mass is attracting all others also suffering from mass. In contrast to Newton's claims, the cosmos shows us that there is no attraction as the formula says $F = G \dfrac{M_1 M_2}{r^2}$ and comets do not fall into the Sun as Newton said it does. You can't have $F = G \dfrac{M_1 M_2}{r^2}$ expressing gravity by attraction and every planet is defying this statement Newton made by having a Universe expanding rapidly every second the clock ticks. Newtonians can't have both worlds with all being true. They can't claim attraction while a Universe is expanding ever since the Big Bang started the expanding! Newtonians can't have a comet claim the Sun is attracting the comet by mass while this is not the case because clearly the comet is not being attracted by the Sun because it moves away from the Sun by the same speed it arrived at the Sun without colliding with the Sun. Yet, Newtonians strive to dispute this truth by having Newton being correct! In defiance of cosmic reality Newtonians stand by the attraction idea and still claim the Universe is in attraction $F = G \dfrac{M_1 M_2}{r^2}$ by the mass thereof. It is not me claiming Newton is incorrect but it is the cosmos rejecting $F = G\dfrac{M_1M_2}{r^2}$ by simply disproving Newton completely. Tel your academic professor in physics to prove that

Newton is correct about $F = G\dfrac{M_1M_2}{r^2}$ and let the professor then show how the comet falls into the Sun with a splash or then if that is not in their ability to prove, then at least have your

professor admit that $F = G\dfrac{M_1M_2}{r^2}$ is just not happening!

Now I am taking my case to the members of the public so that the truth must be brought into the open. I have had the tour they give and then more came my way. I never got around swallowing the mass creating gravity part where science is of the opinion that mass pulls as gravity is… Academics condemned my work and therefore me and for six years where I could not get a publisher to come around and bother to read my work let alone seriously proposing a publishing contract. I had to finally go private with the publishing as all doors shut in my face as soon as the academics read the content of my work because from the nature of my work I take Mainstream science head on and am confrontational on most aspects of astronomy. There does not seem to be any publisher who wants to go head bashing with the establishment of science on official science principles, which I have to do to convey my message in no uncertain language. If you also have doubts about the academics' indisputable correctness, please read on and confront either them or me on everything you read here.

After reading this letter you will have to take sides because you will know the truth.

Then you either become a partner in the crime as you participate in the cover – up of the truth or you will be part of the truth with deceit and help me confront them to acknowledge the truth. Should you think this page is some sort of a prank then answer the following simple question to yourself in utter honesty!

If there is a Big Bang with everything moving apart, how does that support Newton's contraction coming about from gravity being a force of attraction?

How does material repulse material in order to expand while gravity as a force is committed to attraction? Test results received after the Moon landing show the Moon and Earth are moving apart! Yet students learn about mass pulling mass and that puling by mass forces togetherness by contraction.

Newton stated that $F = \dfrac{r^2}{M_1M_2} = F \propto \dfrac{M_1M_2}{r^2}$ which he then immediately passed on as

being the same as $F = \dfrac{M_1M_2}{r^2}$, and that then automatically brought about that $F = \dfrac{r^2}{M_1M_2} =$

$F = \dfrac{M_1M_2}{r^2}$ or as he put it $F \propto \dfrac{M_1M_2}{r^2}$ but held it equal to in terms of being $F = \dfrac{M_1M_2}{r^2}$.

That concept of mass pulling by force was first equated as $F = \dfrac{r^2}{M_1M_2}$.

Mathematically this concept as such is madness because this does not allow for the square of movement where the square used refers to one object moving through space during the square of time. In this, movement is suggested but movement can only be valid in terms of time and with time being absent, movement as a participating factor is not allowed. Let me explain it the best I can…

Movement holds a square value where the moving body starts to move from one point (say v_1) and ends at another point (v_2) producing the square of movement (v^2). However this connects with time or it doesn't connect with anything. If time is absent, the validity of any movement taking place is absent. Time is used to move.

We find a body M_1 and another body M_2 looking at each other across a distance r. Each body is looking from a different perspective across space, hence the r^2. There is no suggested movement in this formula since the square concerns two different bodies at different locations in view of each other where each one is motionless except for feeling a force of magic called gravity. In this formula there is no time connected that will indicate the duration the movement will take in terms of any time unit applying as to when the force of M_1 will destroy its part of r^2 while force M_2 destroys its part of r^2. The connection to movement is in our heads and is not mathematically there as a factor we can use.

The question of validity this formula lacks is it never says how much of the radius r is the mass from either end going to remove and how long will it take the mass to complete this removing...Newton never said that and this missing of time as a component makes this formula invalid. That is why this formula is never used in its entirety forming part of any other formula but only represents a concept...and an invalid concept as well.

In the equation as it reads, it does not allow the bodies any movement say moving from position r_1 to r_2 during time measured at a stationary location...but nothing in the cosmos can be stationary because everything relocates every position in time. You may stand still on Earth, but the Earth moves. While you are seemingly stationary on Earth, the Earth takes you with while moving. The time on Earth is the spin around its axis while this axis is spinning around the Sun and that movement brings about the time we have on Earth. Time is the repositioning of everything in relation to relocating a new position in terms of the Sun.

Time is the movement of the Earth from where the axis was yesterday in relation to the position that the Sun held then going to where the axis is today also in relation to the position that the Sun holds presently and time is the while that the axis took as a day unit to circle and reposition its entire surface completely while putting everything on the surface in relation to a new location on the Sun. Time is the relocating of every point on the surface of the Earth by spinning around the axis while at the same time it is moving this lot around the Sun. I know Newton renounced this when he went completely mad and said $\dfrac{dJ}{dt} = 0$, but this I use to show Newton went bananas when he tried to make this applicable on a cosmic idea. What Newton missed is $\dfrac{dJ}{dt} = 0$ is referring to $\dfrac{dJ}{dt} = 0$ as referring to body spinning that forms the axis around which all bodies rotate and the rotation of the body in relation to its axis brings about gravity.

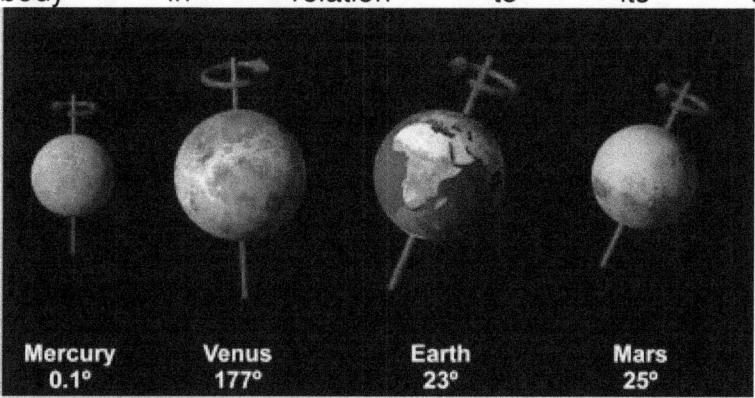

Mercury	Venus	Earth	Mars
0.1°	177°	23°	25°

Time is spinning around the axis while

moving the axis a distance of one day around the axis that the Sun has. That is time and that is what Kepler said time is when Kepler said time is $a^3 = T^2k$. This period of spinning defines the individuality of every planet and allows the differentiation between planets in time.

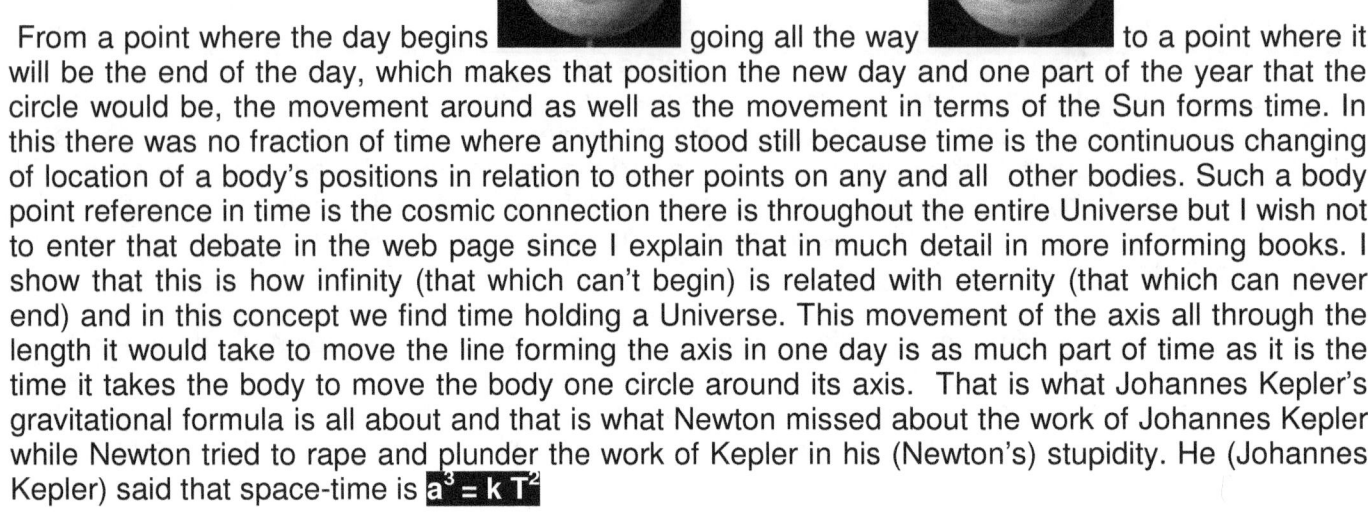

From a point where the day begins going all the way to a point where it will be the end of the day, which makes that position the new day and one part of the year that the circle would be, the movement around as well as the movement in terms of the Sun forms time. In this there was no fraction of time where anything stood still because time is the continuous changing of location of a body's positions in relation to other points on any and all other bodies. Such a body point reference in time is the cosmic connection there is throughout the entire Universe but I wish not to enter that debate in the web page since I explain that in much detail in more informing books. I show that this is how infinity (that which can't begin) is related with eternity (that which can never end) and in this concept we find time holding a Universe. This movement of the axis all through the length it would take to move the line forming the axis in one day is as much part of time as it is the time it takes the body to move the body one circle around its axis. That is what Johannes Kepler's gravitational formula is all about and that is what Newton missed about the work of Johannes Kepler while Newton tried to rape and plunder the work of Kepler in his (Newton's) stupidity. He (Johannes Kepler) said that space-time is $a^3 = k T^2$

The body is a^3. The body has an individual space it holds as well as an individual, space through which it moves in one year. That movement around its axis confines the rotating space from all other space and while rotating, that which rotates with the rotating body is confined with the restraining of mass to that body. Then this confined rotating space specifies identity T^2 in terms of a circle it holds in relation to a circle it spins k going to the Sun and this connects the entire solar system to one point within the Sun. This forms the time per year that the planet has as an identifiable difference. Every planet holds an identifiable space a^3 in terms of time $T^2 k$.

From point A to point B all forms the distance of k. This figure k holds a time/movement relation with the Sun and 365.25 of these forms one year. This is a part of a year that can never be excluded.

The body moves a distance of k around the Sun in one day while during this period of one day...

...the body also at the same time spins T^2 around its axis. That is time...it is the movement of space in relation to one point. Once around holds the measure of 24 hours and one day. This is in ratio with the 365.25 and can never change.

This is time and it involves movement where no point can ever be stationary in relation to any other point in the Universe. The fact that Newton excluded time when he tried to implicate movement of space through space shows very poor understanding of physics on his part.

First Newton decided gravity came to a conclusive value when the mass of one body, the Earth, (M_1) is multiplied by the mass of the other body (M_2) and then the product of this mass multiplication ($M_1 X M_2$) diminishes the square of the radius that held the space between the two bodies.

Newton said there is a body M_1 that is inexplicably connected to a body M_2 across a distance r. That is the entire supposition. This involves Newton's entire vision on what holds the cosmos in place.

The one mass or body M_1 has to be pulling the other mass or body M_2 . Newton went on and multiplied this supposed connection with no clear results at all to indicate how such multiplication will result in movement except by connecting it with a thought planted in my head using culture to provide such a concept. Other than relying on my culture in thought there is no evidence of this taking place! This suggests no motion at all because the square of time is lacking from this argument. Yet, it suggests that this pulling of no motion is going to destroy r from both ends forming the square over space and not the motion of either one towards the other. It never says how long will this movement take and therefore he never confirms a time component connected to such movement. If time is not involved, then movement can't take place. This would mean if there is a force of gravity then the gravity can not bring motion because it does not make room for the time any of the structures would take to move.

Let's concern ourselves with the manner in which Newton saw the gravity apply before he saw such big holes coming into the evidence.

Newton saw an apple falling and in that picture there is clear movement. He saw that the mass of the Earth was pulling the mass of the apple and in doing that, it destroyed the radius.

Having the formula $F = \dfrac{r^2}{M_1 M_2}$ is not the correct way to go about it, because the mass draws the average of one and not the distance becoming $\dfrac{d^2}{M_1 \times M_2} = 0$. If this took place then multiplying the mass would bring about one inclusive factor that the mass would have. $\dfrac{d^2}{M_1 \times M_2 = M_3} = 0$ Then from there mass would be resolved as a unit ($M_3 = 1$) in relation to whatever value d^2 will take on. No one would go to such a presumption when gauging Newton's incredible lack of insight into mathematical possibilities and proof of this is in Newton's statement that $\dfrac{dJ}{dt} = 0$ or that the spin of an object nullifies the spinning ratio. However, concluding that anything can divide in something with a value and result in zero is indicative of very poor mathematical skills. Anything dividing in anything else can never have a conclusive answer of zero. That is a mathematical impossibility and only Newton had the stupidity to suggest that having a ratio (which is what $\dfrac{dJ}{dt}$ has) can be zero. The smallest ratio can be one ($\dfrac{dJ}{dt} = 1$) but never can it be zero...And since Newton did the incredible stupid mistake one cannot put it past him that he thought $\dfrac{d^2}{M_3} = 0$ but having Newton say that

$\dfrac{d^2}{M_3} = 0$ is possible does not make the equation or Newton correct, it simply proves that Newton had no basic mathematical skills to fall back on. It proves that Newton's mathematical development at the time had a lot that could still improve and develop into a better and a more informative insight.

Putting in place that $\dfrac{d^2}{M_1 \times M_2 = M_3} = 0$, this would ultimately insist that mass multiplied $M_3 = 1$ is then giving d a promotional average in relation to mass being one. It is only in this manner that M would be able to influence the value of d in any way. But that is only if mass on both sides would insist on some way to put d in relation to a unified 1 and from that give d an average ratio to mass being one. This does not prove a force of sorts because it proves mass can unify but not bring a conclusive end to the radius between the apple and the Earth. However, as you may guess, the entire result was one big cock-up even Newton could not cheat his way out of. Then he changed

$F = \dfrac{r^2}{M_1 M_2}$ to become $F \; \alpha \; \dfrac{M_1 M_2}{r^2}$, which is equal to $F = \dfrac{M_1 M_2}{r^2}$ and in that,

$F = G \dfrac{M_1 M_2}{r^2}$ that represents gravity and on this inclination Newton founded physics. How misplaced and dreadfully mistaken can so many persons be for such a long time…

The entirety of physics rests on this one formula $F = G \dfrac{M_1 M_2}{r^2}$ after which Newton then still had to cheat even more by locating, producing and manufacturing an invisible Gravitational constant making the formula presented as $F = G \dfrac{M_1 M_2}{r^2}$. However, again Newton plants another motionless force in the midst of cosmology. Again there is a force that holds no movement but only has any asterial of a picturesque value that is forming a part as a force in a fairy tale.

Not one of the formulas that Newton formulated to prove gravity by attraction, be it $F = \dfrac{r^2}{M_1 M_2}$ or

$F \; \alpha \; \dfrac{M_1 M_2}{r^2}$ including the final flop $F = G \dfrac{M_1 M_2}{r^2}$ confirms movement in relation to time and therefore is invalid. As any one can see this formula can't express movement and therefore can't indicate a force of attraction. It is a thought solidified by culture prevailing that cements this idea of moving down and sticking to the ground and it is not derived, confirmed or proven by such a senseless formula such as any of Newton's formula

$F = \dfrac{r^2}{M_1 M_2}$ $F \; \alpha \; \dfrac{M_1 M_2}{r^2}$ $F = G \dfrac{M_1 M_2}{r^2}$ will have us believe. Tell your physic professor to use this formula to show when the Earth is going to collide with the Moon after all the attraction took place. His formula $F = \dfrac{r^2}{M_1 M_2}$ presents no more than indicating a ratio between the top and the bottom. It shows no movement taking place at all and turning the concept around to

$F \; \alpha \; \dfrac{M_1 M_2}{r^2}$ or $F = \dfrac{M_1 M_2}{r^2}$ and even $F = G \dfrac{M_1 M_2}{r^2}$ still doesn't indicate movement in the mass at all!

The attraction theory is without credit even where it starts with the formula that should apply and from there on the cosmos shows attraction by gravity to be laughable. The Hubble constant already disproved this attraction by gravity that should be going on and the Big Bang confirms the dismissal of the Newton attracting gravity supposition! The questions concerning that which you are studying and that which touches every aspect you are academically concerned with, is that if everything is moving apart, as the Hubble constant proves and the Big Bang Theory confirms, how does that evidence support Newton's idea that everything is coming together…and please don't let them fool you with Einstein's Critical Density idea! If there was mass seen or unseen in the Universe and mass generated gravity and gravity does the pulling, then why is the mass not at this moment doing the pulling. What is all that mass of so many supposed stars doing at present while waiting to get to work where it will only later, much later form a force of gravity that then will bring about this pulling of the Universe? What makes the mass slumber in darkness to one day form a pulling force? What has the "darkness" or the fact that we don't see the mass, got to do with the idea that the mass at present is not forming gravity that is forming a pulling force? You are taught that gravity pulls objects to the centre and obviously gravity then has to ultimately pull everything to the centre of the Universe. That is what the Critical Density research that Einstein initiated wishes to establish. The idea is that

$F = G \dfrac{M_1 M_2}{r^2}$ makes the mass create a force that will destroy the radius and ensure everything is

going to come together eventually at one point where the radius then will be no more. If that is the case, then where is that point? If everything is destroying the radius, then it must end at one specific point.

In the classes where you attend a physics lecture, has any one confirmed a location where one might find the centre of the Universe to confirm the ultimate destination of $F = G \dfrac{M_1 M_2}{r^2}$? If you wish to apply a Gravitational constant as a calculated factor then it is apparent that one must know to where such gravity is pulling since it then is the gravity that is predominantly keeping everything apart. Then the gravitational constant is what is resisting the collapse of the Universe. If there is a force, then where is the force taking the pulling…if it is a gravitational constant applying throughout outer space, then where is it having a centre base?

I wrote a book in which I found a means to define gravity. This feat I accomplish and by my effort it was done for the first time ever. For the first time ever runs further back than since the time Newton introduced gravity. Before I achieved that discovery, I firstly had to find the centre of the Universe because it is there that I could locate gravity. I now am able to show how gravity forms because I have detected the centre of the Universe. But by my effort in finding the location I disrupted everything Academics in physics hold holy and for that I am most unwanted in the presence of the Academics charged with guarding the ethics of physics. In short, I clash head on with Newtonian principles. During my research I discovered abnormalities and inconsistencies about mistakes the Arch fathers in physics must be aware of but is hiding with all their considerable influence. I will come to some of the inconsistencies later on but the discovery also introduces a much better vision about many new aspects that I discovered but in reality was never before realised in science. But these discoveries discard and blacken the Newton reputation totally and therefore the academics dispute my work totally in order to save their Newtonian reputation.

The road I took in my search for truth concerning physics was never smooth and the resistance I came across coming from the academic sector was almost unbearable. Academics guarding physics will never allow an outsider to enter their domain and dislodge Newton from being god that is without the intruder paying a heavy price for trying to do so and in this matter I was and still I am seen as being in the role reserved for such an intruder. It is not about my work they detest but it is my rebutting of Newtonian thoughts that they reject! However, such intruding allowed me to find so much that I was not supposed to find. It seemed as if what I found was reserved for all those Super-Educated that studied physics and has now progressed to the inner circles where only the selected few have access to insider information that is available for only a few. The information is not readily available to all and because of that, it was only allotted to the most inner circle and this is some of the insider information I share with you. By finding the centre of the Universe enabled me to find a point the Universe is controlled from. In achieving the locating of the centre of the Universe I had to step on some very important toes, which made me very unpopular. With my unpopularity rating this high, I never qualified for help and those that would help found my ideas intolerable whereby I only found rejection instead of help as I tagged along. Because of this insider rejection I had to resort to private publishing because from the nature of my work I take Mainstream science head on and am confrontational on most aspects of astronomy. This is the only road to go if one wishes to lay axe to the root of the insider corruption they are guilty of. In that sense there does not seem to be any publisher who wants to go head bashing with the Physics Custodian establishment of science on official science principles, which I have to do to convey my message in no uncertain language. I argue that if it is the correct practise to use $F = G\dfrac{M_1 M_2}{r^2}$ to calculate gravity, then the radius holding the gravitational constant must lead one to the centre of the Universe. With nobody willing to publish my work as I confront science dogma and principles all the way, I had to go the road alone and fight the battle by my private effort.

This is only one of many points that I make on this one issue and there are so many other issues one may think of those in terms of counting in numbers in many hundreds or even in thousands. If the Sun for instance has mass that is apart from the Earth and the Earth also has mass and there is a gravitational constant in between the Sun's mass and the Earth's mass, we have the radius in that location. It then must be the gravitational constant that fills the space that the radius holds. It is rather obvious that while the radius is filling the vacant space between the Sun and the Earth, it is the only place left where the gravitational constant can hide. To find the centre of the Universe I had only to find the gravitational constant that holds the centre. Through my venture I discovered one person that knows what gravity is! Newtonians went and filled that space reserved for the gravitational constant having a measured value, with nothing! How can nothing have a value of 6.67×10^{-11} while also being filled with nothing as it is nothing filling the nothing of outer space?

If you think scientists know what gravity is, do not be duped that easily because no one in science remotely knows what gravity is…not even Newton knew what gravity is except Kepler… and because of what Kepler introduced, I now know I can prove what gravity is. Gravity is precisely what Kepler said gravity is and only Kepler knew where to find the centre of the Universe because only Kepler knew what gravity is all about.

Try to get an answer from any academic person in physics about where the centre of the Universe is, is like trying to touch the moon.

Again I repeat the following by which I end this chapter as a concluding thought: I dispute Newton and so should all students that is learning physics because Newton's arguments are an onslaught on human intellect. Think of the resentment that students have towards Newton under normal conditions when they have to cope with understanding the Newton principles Mainstream science says are applying and how that confusion of what is possible and what Newton suggests is possible clashes with their intellect which makes them feel stupid. Students hate Newton because they don't understand Newton and for that they are accused of not having the intellectual capacity to follow Newton. Every student from the past going into the present and even including those forming a future generation of students will purchase a book that is showing that Newton's legitimacy is cracking up when exposed to some vivid scrutiny. This fact gives the book a selling potential like no other book in the past could do. Yet, I am unable to find a publisher because publishers need academics to assure the correctness of the information in the book and academics would cover up Newton's errors at all costs. There is a total denial about the truth and as long as those academics have the opportunity to brainwash students in to accepting Newton's unproven and ridiculous concepts and as long as their misconduct of mind control by fact manipulation goes unchallenged, the process will go from generation to generation as it has been going for the past three hundred and fifty years.

In short I will now explain what I explain throughout the book you are about to receive and which is named *Newton's Fraud* or whatever it will be named as. The Newtonian formula $F = G \dfrac{M_1 M_2}{r^2}$ is the formula used by science to explain and define gravity. It says the that the ($\underline{\textbf{M}_1} \times \underline{\textbf{M}_2}$) mass of one object pulls the mass of another object and this process in relation with a gravitational constant (**G**) (a supposed force keeping the universe attached) and the pulling subsequently destroys the radius (**r²**) being between the objects. That says that objects **ALWAYS** <u>**MOVE CLOSER**</u> *BY FORCE* in relation to **<u>MASS</u>**. Newton submitted the suggestion that objects fall as MASS provides the force that will cause the falling by the inducing of a force he named gravity which he subsequently only proposed was the acting suppositious force. I disprove this formula in so many ways in this book and I show that this formula and the ideas Newton introduced just don't stand up to even the smallest tests. Then, if Newton's idea on gravity has validity and mass is responsible for objects falling, then all objects that are in a process of falling must be subject to mass and in that idea rests differentiation and discrimination in size and compactness producing speed variations.

If any and all falling is subject to the variation mass introduces and the influences coming about is the result of mass interfering in the gravity force being generated, this then must bring different speeds to cause substantial variation in the falling of different objects holding different mass factors. There can't be conformity in the falling of all objects while such falling is the result of the discrepancy that mass has to inflict due to variations that result in mass differentiations. This is a vital issue that science eludes and has all clever ways to avoid direct questioning. This part science just run around and never addresses and avoids confronting the issue. This avoidance of confronting the issue whish will disprove the validity of Newton is done with such cunning as you will not believe. The fact that objects fall due to conformity in the falling, science accepts but portrays a picture of deceit that mass brings falling distinction and therefore equal falling doesn't happen, while they at the same time admit to Galileo's presentation that falling of all objects are equal in tempo, irrespective of size or any form of differentiation. While they promote the obscurity that Newton and Galileo is in harmony the truth about their deceit is that the two can never have the same issues. That I prove is a fact and also I show how big a part this is in the overall covering up of Newton's initial fraud.

Let anyone reading this disprove any point that I made this far!

Module 3
<u>Glued or Not Glued</u>

Do you know what gravity is?

I ask this question because Einstein couldn't answer it. I ask this question because I have serious doubts about the fact that there is gravity anywhere. I have doubts as to the realisation in physics about the authenticity there is formulating the concept applying as a force called gravity. Hearing this, everyone in hearing range steps back in shock and disgust because everybody suddenly suspects my mental status. Everyone thinks my state of mind could be infectious and contagious. In every book I read and in everything I studied I never could find any trace of gravity. How does mass bring about gravity? What is mass that it finds the ability to employ something as vague as gravity to pull and pull? Those well to do Masters of physics already gave the undiscovered particle that does not truly exist a name. It is going by the name of graviton but that too is matching the same principle as the Phantom (and his white dog or is it a white horse?), Tarzan, Superman and a host of aliens not yet introduced to man, but in retrospect when all are considered, the idea is foolish, which by any standard is not entirely scientific.

I have serious doubts about gravity being present. To my thinking I concluded that there is no such a thing as gravity. If gravity was present, we should find gravity and we should be able to define it much better than it being a Neanderthal concept of a force presenting little understood dynamics. That gave me a goal and a direction in which to search. Ten years later I was still convinced there was no gravity because no one had found gravity. Then twenty or so years after my introduction to Cosmos and Carl Sagan I was reading one night about Einstein being of the opinion that he thought if he fell out of the patent office in Austria which was a multi story building, that he would then experience the feeling as if he (Einstein) would be without weight while descending to the ground. He realised that if he, the pen he had, the table he was next to and the chair he was sitting on would fall through the window, then the lot would fall together. While descending, all were then in a free fall and it would seem as if the lot that was falling would have the same mass because they all were falling in the same manner. For the first time in my life I had an issue about something that Einstein said. If he felt that he was without weight it could only be because he was truly physically without weight and not because his imagination was running wild. There was no middle ground of having mass and imagining being without mass. That means I am questioning gravity. If they are falling equal, then there is no gravity pulling them differently. That means the gravity Newton saw is in Newton's imagination and not in Einstein's imagination. However, that is not the way I am perceived, because since the first time I uttered the notion, I was considered by friend and foe as fulfilling the role of the village fool. Still, I am of the opinion that there is no gravity…mad as that may sound.

If I say there is no gravity I am not trying to convince you that you're being on the ground is entirely just your imagination holding you on the ground. Your being on the ground is not due to deeply rooted physiological issues following you in the form of a concealed depression since your childhood and is reminisce of a dark period from your early days that you never got to terms with. It is also not due to your willpower or maybe your lack thereof that is preventing you from flying to Mars. Hey, I am not that mad… I wish to give you a test to judge your intellect seeing that you (and almost everyone else on Earth) think of me as filling the role as the village fool, since I am in doubt about gravity and the being of gravity presented as a reality.

Do you know what gravity is?

When hearing the question I put to you, you immediately jump to the conclusion that I refer to that which holds you steadfast on Earth. I am referring to gravity and you know what gravity is. It is gravity that holds you on the surface of the Earth, where you've been stuck ever since you can remember and the sticking goes on relentlessly. It is what you have been fighting since birth and if not for gravity, you would be a Superman. The only thing Superman has that you seemingly, obviously and definitely don't have, is his Superman ability not to be restrained by gravity and confined to the ground. If it was not for the effect gravity had on you, then you could have been the local superhero along with six billion other Superheroes that had no restraint from gravity. By your suffering from gravity during all your life, that puts you in terms of being an expert on the subject of gravity and no one knows gravity better than you do. The only aspect of your life you are unable to change, is your attachment to the restraint of gravity.

To your knowledge there is only one form of gravity and you know better than most that you are standing in aide of that gravity, where it is that gravity that is keeping you on Earth, so by experiencing the restraining it brings with the concept for your entire life and during your entire life, you very well know what gravity is. Your expertise on gravity had you fighting gravity more than anything you have ever fought or had any other fight with, including your Mother-in-law. If it wasn't for gravity, you could out-accelerate a fighter jet in mid air. You know it is gravity that is going to get you old and it is the very same gravity that is telling you that you now desperately have to lose weight or over strain your heart and die young … then I come and accuse you of not knowing what gravity is by asking you what gravity is! You are so sure about gravity that you are no more familiar with any other subject and little else has your expertise as gravity does.

Are you sure that gravity is what you might think gravity is and that you and Newton have the same concept about what gravity is?

Have you ever thought that which you think of, as being gravity, is not that which science presents as gravity? The idea you have about gravity is not the gravity science says is gravity. You might think that your being stuck on Earth is that which you think of as gravity...that is the concept you formed ever since your first attempt to sit up straight and that was just after you formed an opinion about the benefits of milk. Since then you needed no one to inform you about gravity because since then you have a very clear idea of what you envisage or what you think personifies gravity. I say that if what you think of gravity as that which is keeping you glued to the Earth, then it is not the gravity, science defined as gravity. That which Science defines as gravity and that which you are thinking of as gravity is very much not equal but also it serves the Masters in physics well to leave you with having that idea about your gravity and their definition about gravity being the same because now they don't have to inform you what gravity really is. They leave you with what you think of as gravity. You now are so well prepared as to become a candidate to be brainwashed in believing your way of believing is their way of believing what gravity is and in believing you share their view about gravity, you are in the best prepared state to become another one of their millions they have mind control over. What Science says gravity is, is far from your view because the Newtonian's approach to gravity is the same as a magnet hooking onto a metal. The grip coming from that is gravity, according to science. They say there is a force between you and the Earth and this force is pulling you as much as it is pulling the Earth, but it is pulling the Earth much harder than it is pulling you because the Earth has much more mass than you have. That means there is a force within you and there is a force within the Earth and these forces coming from within pull together that, which should be apart. That form of gravity will have you locked onto to the Earth with no release but when you do have release then the release will allow you to escape. The closer you are to the Earth, the more significant and powerful the force then must be. The worst part of breaking the force, is the first millimetre and from then on the rest is child's play. That is what they suggest when they say mass is pulling mass to reduce the radius in $F \; \alpha \; \dfrac{M_1 M_2}{r^2}$, which is $F \; = \; \dfrac{r^2}{M_1 M_2}$, which later was changed to $F \; = \; \dfrac{r^2}{M_1 M_2}$. What this means is if your mass is hundred kilograms and the Earth has a mass of 5.974×10^{24} then when you are standing on the ground with an infinitely small radius between you and the Earth, then the force keeping you attached to the earth is eternally big. With the radius of say one billionth of one millimetre you have no chance to be released from the Earth, while when you are say one kilometre way from the Earth the force is one thousand million billion times weaker. That is shit at its best because notwithstanding height, the gravity remains the same. $F \; = \; G \dfrac{M_1 M_2}{r^2}$, $= M_1 = 5.974 \times 10^{24}$ and M_2 = 100 kg divided by the incredibly small radius of 1×10^{-15} m then the force gets to be $1 \times 10^{+15}$ making the mass more proficient by a margin of $1 \times 10^{+15}$ and that is totally ridiculous in reality but if the formula is correct that is what should mathematically be true. With a magnet, where this concept is true, the last millimetre makes the magnetic pull so strong it becomes almost humanly impossible to control the distance between the two magnets, whereas when there is a meter distance between the two magnets, there is no pulling power to speak of. The same should apply in the formula if the formula $F \; = \; \dfrac{r^2}{M_1 M_2}$ did apply. Once you are free from the ground, your first stop should be the moon, because then you overcame gravity at its worst. We know this is not true because jumping the first meter is the easy part. Getting higher than one meter becomes more tiring and from two to three meters a person needs other devises aiding the jump. The further the jump, is the harder the task is but contrasting this formula $F \; = \; \dfrac{r^2}{M_1 M_2}$ would suggest that the smaller the radius is, the harder the effort must be to break the gravity strangle hold.

$$F \quad \alpha \quad \frac{M_1 M_2}{r^2}$$

When the radius is insignificant, the mass becomes enormous. The opposite also applies because when the radius becomes enormous the mass and therefore the force become insignificant. That is just the way it works when there is such a directly suggested relevance applying between the mass bringing on the force in relation to the radius influencing the force.

$$F \quad \alpha \quad \frac{M_1 M_2}{r^2}$$

With the radius big, the influence of mass and the force is weak. This would then suggest that if we are able to break the first meter between the Earth and us, it is then the end of confinement as we would be able to fly to the Moon for weekend shopping on a Saturday morning. We know that is not true because if that were true, walking on Earth would be desperately strenuous. It is getting into the air at an increasingly higher distance that presents the problem. What we seem to experience is like a blanket of something pushing us down and the first lift is the easiest. It seems as if the strenuous pushing starts when we try to lift the blanket of air way up into the air. Partially lifting is not the problem but breaking the cover altogether serves as the real problem.

Now, you might say Newton did some damage control when he changed the formula from $F = \frac{r^2}{M_1 M_2}$ to become $F = G \frac{M_1 M_2}{r^2}$ because Newton saw some controversy in the way I explained the working principles of the formula. Don't you believe that one because Newton made the principle even less applying and much more complicated!

The changing of the formula from $F = \frac{r^2}{M_1 M_2}$ to become the formula $F = G \frac{M_1 M_2}{r^2}$ was done in order to give the concept a cosmic significance. When using the formula in terms of $F = \frac{r^2}{M_1 M_2}$, it only refers to an apple falling from a tree to the ground. There is no sense of dignifying the true nature of the event by supporting the refulgence that would reflect upon the appreciation of the spectacle Newton witnessed the day from which such splendid scientific demiurge eventuality arose, where $F = G \frac{M_1 M_2}{r^2}$ indicates gravitational implications of a cosmic nature. Using the formula $F = G \frac{M_1 M_2}{r^2}$ places Newton's lustrous occasion in appropriate perspective.

What the grounds were for making it a cosmic notion still eludes me to this day…because after all was said and done, it was only an apple that fell from a tree as the formula $F = \frac{r^2}{M_1 M_2}$ would suggest, and not the Moon coming from space and landing near Newton as the formula $F = G \frac{M_1 M_2}{r^2}$ would suggest, but I am getting to that later on in the book. The minute the radius r disappears, the mass is overbearing and the minute you find the means to break the strangle hold of mass keeping you on Earth; it is like a magnet releasing with no effort available to secure your position again. In the case of gravity being the same as a magnate, it will mean that an object will fall faster and indefinitely increase the descending rate as the object falls, since the object's mass is increasing in force by the diminishing of the reducing radius. Such reducing of the radius will increase the strength of the mass.

An object will fall while descending onwards with limitless acceleration and in a fall the object should even be able to break the sound barrier, if the object is massive enough or if the object was dropped from far enough, but this Newtonian lie is far from the truth.

This concept proves that it is movement more than mass ever can, that features prominence in gravity. Even with an object as massive as a 747 Boeing aeroplane, the aeroplane of such size does not require much higher speed than do a fairly smaller aircraft in order to achieve take off speed. It is the relevant size that the wings carry in relation to the space the aircraft move through during a specific time period that determines the flight height required to maintain flight space. If the formula $F = G \dfrac{M_1 M_2}{r^2}$ did apply, the biggest problem to overcome would be to be able to break the hold the mass would require in that first millimetre of lifting. To be free from the Earth, the toughest part would be launching the escape of the object in the first few millimetres of lift, after which the task would exceedingly become less problematic as the distance between the Earth and the mass grows.

This Newton formula $F = G \dfrac{M_1 M_2}{r^2}$ works on the principle through which we observe. When the radius is small the object we look at will seem big and when the radius between the viewer and the object is large, then the object will seem small. We know that that is how vision and visibility works but it is hardly the method in which gravity works. Looking at flying, we find it takes more effort to lift an aeroplane higher and keep it high in the air. This is the opposite of what we can deduct from using Newton's formula where a stronger distance will bring about a much weaker force.

They give you one concept while their approach has quite another. Using the formula $F = \dfrac{r^2}{M_1 M_2}$ or in its cosmic interpretation being $F = G \dfrac{M_1 M_2}{r^2}$, connects your position on Earth in gravity in the same manner as that which the magnet is applying.

This is mind control just like UFO's are used to culture the very same principle, which all the institutions from the Government to the press and all the Academics at NASA and in agents at NSA, taxed with suppressing information, are benefiting from the misinformation because that suits their use in mind control on you being the victim. They use it to hide behind while they dictate the terms by brainwashing the people into reluctance. The entirety is very deeply rooted in society but in this book I only aim to tackle gravity and therefore we return to gravity.

We have to analyse what gravity is. There is a defining to do by dissecting the factors we see as gravity. Gravity is not many things to lots of people, but is very specific one thing. To investigate the identity of gravity, there are questions one must first answer. Is gravity the part that has me standing on the Earth or is gravity the part that has me glued to the Earth. It is surprisingly not the same thing. While I am standing on Earth I am standing still as far as my perception carries but there remains the intention to move, even with me standing still. I can't stand still and at the same time intend to move by the same driving effort. My standing on Earth but remaining in intention to move further can't be motivated by a similar or the same principle. The concept is not the same because when standing still, the Earth does the moving on my behalf and there is no independent motion of my body. But when I jump, I am moving away from the Earth and when I walk, I walk along the surface of the Earth giving me independent movement on top and in addition to the movement of the Earth. My movement is stopped by mass and not committed by mass when in mass. When in mass, my intention is to be motionless. There are two occurrences where one is immobility through mass and the other is being without mass while moving. Mainstream science wants you to believe it is the same thing, but it is not. Gravity is the inclination that my body has to move further towards the centre of the Earth while mass is, that, which frustrates my body, and prevents it from moving towards the Earth. Science tells you it is the same thing but it is not.

The very second that that which prevents me from moving falls away, the inclination of moving, which is gravity, kicks in and gravity lets me move freely again. I will continue to move towards the centre of the Earth and that process has a very well defined and scientific term which we use to describe the event: We call it falling in English but every other language devised another word to describe the same concept. When I stop falling that which gives me mass is what ends my moving. What gives me mass changes my gravity from movement to being inclined to move. As soon as that which prevents me from moving allows me to continue to move to the centre of the Earth, my gravity will start giving me full motion and then that which gives me mass by frustrating my independent movement will not stop my progress towards the centre of the Earth. When my gravity is stopped by mass blocking my movement and that, which gives me mass, then has the ability to turn my descending towards the Earth over to a frustrating of gravity or my freedom to move and change such a freedom to move into a tendency or an attempt to move. The price I pay for the loss of free movement is the gain of mass in mass preventing further movement. It is not as simple as being glued or not being glued…

To the average people, the idea that I am attacking what they accept as forming gravity is unacceptable. When I discard this idea they have about gravity, they see it as me rejecting that which they accept as being their own. The gravity they understand is also what they grew accustomed to from childhood. Gravity is what makes you sticky and what holds you tied to the ground. When I would remark that there is no gravity and with that I require the person to expand his or her mental view about the issue of gravity, I am met with aggressive animosity. They associate Newton's gravity with what they experience as if Newton said gravity is what they view as the force that is keeping them glued to the Earth. That idea that gravity is keeping everyone locked to the Earth is normally achieved by Newtonians that are promoting the thought behind what we experience in our daily life as gravity and when I dismiss the idea normally accepted as gravity, this fact becomes overwhelming overloaded in their minds, which leads to confusion on their part in understanding either what I try to convey or what they understand Newton said what gravity is and with that they experience my defying their view on gravity as me aggressively trying to harm their sanity by dangerously attacking their mental stability. Newtonians promote gravity in the sense that it becomes a slogan. Everyone uses slogans that bring comfort to all persons as the slogan covers the concept and when I deny them their view about gravity, I am removing the concept by taking out the comfort zone. I am removing the safety people experience with forming huge concepts while receiving very limited information. This is today standard practise in common use and this is a process with which all persons are brainwashed. On all fronts this is accepted as a process used everyday as part of the advertising intellectual age.

If I say gravity is the part not connected to mass that brings on the intention of motion to carry on moving downwards notwithstanding the blocking action which comes from intervention of space occupying by a controlling body while mass is not connected to gravity, since it is stopping the motion of moving downwards, the concept starts to override the normal commonsense. Mass is having a much more demanding space filled with material in a position that will intervene further movement of descending to a centre of the body having gravity and therefore performing the descending motion. Then my threatening to burst this super intellectual "two-word in one idea"-slogan mentality making everyone a genius, most people see as culture threatening and in the defending of that superior feeling, they only see that I wish to destroy them and challenge me to protect their comfort zone.

When I state that there is no gravity they all respond in a manner that is putting me under suspicion and not one person responds positively by asking me why I put gravity under suspicion. Now I give the reasons why I challenge the concept of gravity and I would love to see a person challenge my challenge. This is why I discard the fact of gravity as promoted by the Brainy Bunch.

By definition gravity is defined as being:
Gravitation is the *force* of *attraction* that *operates between all bodies*. *The size of the attraction* depends on *the masses of the bodies* and *the distance between the* : *the gravitational force*

diminishes with the square of the distance apart according to the inverse square law. Gravitation is the weakest of the four forces. Newton formulated the law of gravitational attraction and showed that gravitationally a body behaves as though all its mass were concentrated at its centre. Hence the gravitational acts along a line joining the centres of the gravity of the two masses.

It is not you being glued or not being glued to the Earth that I discard. It is the definition holding this whole idea that I do not share in the least. What the definition describes is magnets pulling and it is the total opposite of what I experience. Breaking the first millimetre of gravity clampdown is the easiest and not the most difficult. The difficulty increases as the radius grows but gravity does not reduce as the radius decreases where $F = G \dfrac{M_1 M_2}{r^2}$ would suggest that this takes place.

When I say there is no gravity, everyone thinks I say we all are going to fall off the Earth at random and with me thinking that way, then it is obvious that I must be a nut. Everyone thinks of me as the clown acting mad when I say gravity is not to be found in nature. But I do not say we are not standing on the Earth. I do not say there is nothing that is keeping me glued to the Earth. I say there is no attraction between two bodies by the force of the mass that in such doing is diminishing the radius parting the bodies by the inverse square law. I say there is a connection by motion between the centre of the body and the material surrounding the centre. This is what I say when I say there is no gravity.

I dispute Newton and so do all students at first when students are forced to learn about Newtonian physics because Newton's arguments are an onslaught on human intellect. Think of the resentment that students have towards Newton under normal conditions when they have to cope with understanding the Newton principles Mainstream science says are applying and how that confusion of what is possible and what Newton suggests is possible clashes with their intellect which makes them feel stupid. Students hate Newton because they don't understand Newton and for that they are accused of not having the intellectual capacity to follow Newton. Every student from the past going into the present and even including those forming a future generation of students will purchase a book that is showing that Newton's legitimacy is cracking up when exposed to some vivid scrutiny.

I have written several books in which I challenge the thought process of Mainstream physics and especially Sir Isaac Newton's arguments about physics. I am of the opinion that even though everyone thinks of Sir Isaac Newton as the genius who established every aspect that is used in modern physics today, but in spite of every other person hailing Newton, I remain of the opinion that the man did not have a foggy clue about any of the principles driving the concept that he named as gravity, or what brought about gravity according to his explaining of what forms gravity. I am able to explain gravity but it doesn't even vaguely resemble Newton's version of gravity. I can explain gravity by proving my explanation with the use of simple mathematics. I use Johannes Kepler's formula to back up my statements. By using Johannes Kepler's formula I found a way to prove there are four phenomena found in the cosmos. There are the four phenomena applying in tandem that together form gravity. They are: **The Titius Bode law**; **The Roche Limit**; **The Lagrangian Point System** and; **The Coanda effect**. As the phenomena don't support Newton's vision on cosmology, the phenomena has no support amongst Mainstream science although they did apply it with a positive results in locating the missing planets at the time of their discovery. When they located unknown and undetected planets in the past, the existing of the phenomena was never disputed, but when the argument of proving them comes to mind, then they are dismissed as some coincidental abnormality occurring. But since it holds no similarity to Newton's view on science, Mainstream science rather disclaimed the validity of the phenomena than they would find fault with Newton's ideas. In the mind of science the cosmos can be wrong and God can be wrong but Newton can never be wrong. In using the four correct principles correctly, which I back up with the correct mathematical interpretation thereof in support of the function that each phenomena has in forming gravity, I did a far better job than what Sir Isaac Newton did and what I achieved is of a far more acceptable level as well as being mathematically far more correct than what Sir Isaac Newton did achieve with his guessing about issues he couldn't explain. To be successful in my quest to find an explanation for gravity, I had to redirect all my concepts I previously had and also alter all the otherwise normally accepted thinking on physics. I had to find the phenomena and I had to dissect the function of each phenomenon as well as mathematically valuate the phenomena. In this process I realised that to come to realise what gravity is, I had to realise that gravity is not what Newton saw forms gravity.

Newton devised a formula $F = \dfrac{r^2}{M_1 M_2}$ that represented gravity. Newton thought the mass of the apple that fell, drew the Earth as much as the Earth drew the apple. The one mass factor represented the mass of the apple while the other mass factor represented the Earth and the radius was in place of the distance that the apple had to travel as the apple fell from the tree in view of Newton. This falling he saw as the gravity that the Earth's mass and that apple's mass were achieving. Let us have a look at the force F that Newton introduced.

What is F and what worth has F while we find out what role F plays. Let's place F in $F = \dfrac{r^2}{M_1 M_2}$ and find what F really has in a mathematical sense.

$F = \dfrac{r^2}{M_1 M_2}$ can be replaced by $F = \dfrac{a^2}{a_1 \times a_2}$ which then would leave $F = \dfrac{a^2}{a^2}$ that leaves **F = a^0** and that outs the factor of gravity without value or worth being a factor of **F = 1**. This doesn't make much sense but Newton never saw this imperfect outcome to his otherwise perfect formula. But calculating $F = \dfrac{r^2}{M_1 M_2}$ in terms of real factors worth, it makes no sense in another sense.

Replace all the factor values in terms of $F = \dfrac{r^2}{M_1 M_2}$ and with the mass of the Earth multiplied by the mass of the falling body which was the apple, the force is exceptionally small. If I calculate the force in terms of my view, the force that comes about in this formula the result we find in calculating Earth with the apple divided by the distance between the two is something in the region of less than what the mass of one atom would be. This left our genius with some headache and a large problem (or is it a very small force of gravity) to solve. The force coming from this equation is less than microscopic small! Then Newton improvised masterly by cheating the wits out of all mathematical logic.

When they use another formula that also uses the symbol **F** in the formula **F = mv** I still have to find one academic who can show me whereto did the mass of the Earth disappear while taking with it the gravitational constant as well as the diameter parting the mass **m** from the other disappeared factors. This is one of the many small issues they never think of because they can't explain it while upholding the correctness of Newton at the same time. Let one of them with the many doctoral degrees, show how they come from $F = G\dfrac{M_1 M_2}{r^2} = F\; \alpha\; \dfrac{M_1 M_2}{r_2} = F = \dfrac{r^2}{M_1 M_2}$ to eventually reappear on the surface as the formula **F = mv**. If you thought gravity was an act of magic, try this magic. Where did all the factors (M_1, G and r^2) go while being on route to change in appearance to become **F = mv**. The mass of the Earth that academics in physics claim is there and which is supposedly doing the gravity pulling, is a relevance that the object has with the Earth having a factor of 1 and this relation is effective viable only when the object having this mass is resting on the surface of the Earth or having some direct contact through another medium connecting the object to the Earth. The object rests on the link by a link or otherwise is resting directly on the Earth, but the condition of mass that any object has, is that the object is standing still or moving while being in direct contact with the Earth. But all action the object has is relevant to the position the object has in relation to an allocated relevance with the Earth and relating to the movement that the Earth has. The object in mass has to move directly with the Earth or slightly more than the Earth. The object only shows having mass when connected and when accepting the movement the Earth has, but the mass the Earth should place into the calculation alongside the mass the object has, that as a complimenting factor is totally absent in normally used physics because the Earth has no mass. The Earth's mass is lacking all visible presence in influencing physics by lending support or increase any calculation in physics. This proves my statement that it is because the Earth and all other planets do not have mass and therefore can't be used as a calculating factor.

Planets have no mass and neither has the Sun got mass except the mass Newtonians wish to credit planets with. Bigger planets don't move faster because they have more mass and smaller planets are not further from the Sun because they have lesser mass. All planets big and small spin at the same speed around the Sun and in relation to the Sun and all planets are scattered going around the Sun while being big and small where all sizes are well mixed. This is because planets have no mass except in the imagination of Newton and his devoted followers. The mass of the Earth never plays a role in physics and the mass of planets do not draw any of the planets closer to the Sun and let one physics professor bring proof that the planets do draw nearer to the Sun!

They just can't because planets do not have mass that can produce a pulling gravity! If and when the mass of the Earth do not feature as a factor in any formula that is used in physics, then the mass of the Earth is no factor playing part in gravity. This then can only indicate that the Earth has no mass. If there is an absence of mass as a factor that influences physics, this can only be as the result that the Earth mass has no gravitational presence in any physics formula. Gravity does have the value of $g = 9.81\ Nm/s^2$ but that I explain and the value $g = 9.81\ Nm/s^2$ I prove as well. With that evidence being that clear, then the mass that the Earth should supposedly have, does not produce gravity as Newton suggested. Prove me wrong by getting gravity at $g = 9.81\ Nm/s^2$ from using either of any of Newton's formulas being $F = G\dfrac{M_1 M_2}{r^2}$ or $F\ \alpha\ \dfrac{M_1 M_2}{r_2}$ and $F = \dfrac{r^2}{M_1 M_2}$. Let me see Newtonians do that and I will become a believer in Newton! The Earth has no mass because physics can't show the Earth's mass playing part in calculating formulas and if there is no mass that plays a part that should produce gravity, then mass can't be responsible for the producing of gravity as Newton declared. That makes Newton's suppositions total rubbish and that makes Newton responsible for a crime of defrauding and falsifying the science of physics. If you, the reader is able to get academics in physics as far as even reading this argument I make, then you are more influential than I can ever be. They plainly dismiss all these arguments with arrogance by discrediting my credentials!

What Newton saw as gravity can't withstand even the slightest test of proof and I showed that it is not possible to use Newton's formula as Newton suggested it applies to mathematically calculate gravity. I come back to this issue later on. I have tested Newton's thinking and the book I offer to you for investigation serves as the testimony to all the testing I did on Newton. This any body who can see, will see when reading this book, I tested Newton from all the angles to see if he possibly could be correct but found his thinking wanting every time. The truth about Sir Isaac Newton's concepts I came to conclude, was that the reality is that it is not in any way overstated to declare that Newton conspired to defraud science and moreover that he committed blatant mathematical corruption in trying to prove the concept he had about what he thought forms gravity. There is no backing for Newton's ideas and even the ideas which are in use are not in the form that Newton said it applies where physics in daily use serves as the best discredit to Newton bringing no proof about any of the claims that Newton made on matters concerning science in cosmic gravity.

The phenomena that I use is still to this day unexplained by Mainstream science because it shows no sign of using mass and without mass the Newtonian mind understands nothing!. Newtonians don't understand the four phenomena due to the fact that science up to the present date has no means or method to explain the four mentioned phenomena while I can explain the working of each independently and how they work in a combination to produce gravity. I found a way to put those four phenomena in a perspective and put the four in a mathematical sequence that from there I could explain gravity in detail. When I first approached academics, I had the opinion that all academics were knowledgeable about the lack in the correctness we find in Newton's views and that every one in physics would be rejoicing in finding what gravity consists of. I was under the impression that I would be embraced by those in physics for finding a solution to Newton's errors. I was in for a nasty shock with such naivety.

I met with such rejection that no one even cared to look at my work because they were of the opinion that looking at my work would be sacrilegious to Newton. I was told on occasions that Newton has never been proven incorrect and therefore any attempt on my part in doing so is a waste of time. At first I was not confrontational towards Academics in physics and avoided any indication about disagreeing with Newton, but academics always threw Newton at me and eventually for self protection I had to start to confront them and confront Newton, with whom I was in disagreement from the beginning although at first I was reluctant to voice any opinion about the matter. But slowly it dawned on me that if I had any serious plans to introduce my ideas, I had to dispute Newton's gravity principles and show the inconsistencies and dishonesty in Newton's approach to physics. I came to realise that his flaws are there and the mistakes are present whether I avoid it or attack it; the inconsistencies are part of forming the basis for modern accepted science. It is that strangle hold I had to break before I could even think of finding acceptance about change.

Then slowly I concluded that only and after I can get people to see how incorrect Newton is, do I stand any chance to introduce my line of thought on gravity and I am so sure of my ideas being correct that I dare any one to disprove any part or the entirety that forms gravity as I see gravity! But that can only come about when I can get an audience to see how I expose Newton for what Newton was and that is where I find no luck. I can't find one academic with influence who is brave enough to stand up and face my attack on Newton and argue me down or prove me wrong in a sound debate. The moment any academic realises he or she is reading my condemnation about Newton's correctness, their minds shut down! No other thought can penetrate their mind but to think in terms of Newton being correct even when confronted with facts proving Newton incorrect. They stop reading my work. They do not get confrontational but defensive and in defending Newton they refuse to read further!

I realise that everyone has the view that my finding fault with Newton shows signs of madness and progressive signs of dementia on my part and in my thinking to even regard any possibility that I am the only person on Earth that is correct and all others that ever studied physics are wrong, is pure foolishness, but mad as it seems, if that is what I have to say to be correct in what I say, then that is what I say...Newton is wrong about gravity. I don't say this lightly or without understanding the enormity of what I suggest is going on, but be that as it may seem, it is the truth without question that Newton went on for three hundred and fifty years defrauding science with no one testing his claims.

Detecting Newton's misconduct is possible because I saw a way to break away from the invalid concepts Mainstream physics holds. I saw where Newton went wrong and correcting the major mathematical error is so small...if only any one would listen! ...And of course one has to admit that the Earth doesn't pull or push by mass or any other way just like the physics formula they use to formulate indicates. Notwithstanding the pose Mainstream physics tries to uphold, the entirety of physics still uses the idea of magical forces intervening in nature and they still base concepts on unexplained novelties. Think of how they found four unexplained forces going around and influencing persons in an unexplainable manner except that they can see that it is through the inexplicable magic of gravity keeping people attracted to the Earth. To say the least, the concepts physics use in terms of Newton would not even be acceptable to children in the modern informed era we live in. I challenge any person to prove Newton, not to accept Newton but to undoubtedly prove Newton correct! I recognised the impossible double standards Mainstream physics apply to promote their much shady explaining. In short I tested Newton's principles and found the principles to be wanting on all levels of consideration.

The statements that Newton introduced have inconsistencies and to cover these holes science has in their understanding of cosmic principles, they have to apply standards which are symptomatic of having total confusion. To compensate for these bogus truths that were supporting their incredible theories, they simplify issues to such a level where what they embark on is equal to and the same as witchcraft and soothsaying. They admit the cosmos is expanding but the expanding is not here in our neighbourhood because in our neighbourhood Newton rules and therefore notwithstanding a Big Bang still applying and a Hubble expanding going on, in our neighbourhood we contract because Newton said so and Newton just can't be wrong! Their pitiful explaining of the fundamental working of physics is meaningless. In spite of finding evidence that the Moon and the Earth is growing apart in distance, they still uphold Newton's view on contracting because it suits their work and leave everyone under the impression that the contracting is valid as it should be if Newton is correct. The Earth and the Moon is growing apart at the same rate as human hair and human nails grow. Because they lack true basic understanding, they have to accept the unproven and it remains unproven that the cosmos is coming together by the power of mass that is inflicting gravity.

Do you think of astrophysics as being the department that is run by the wise and the level minded, the honest and pure at heart, the nobility of well-to-do academics and the sober thinking standing in front of the world as the absolute trustworthy? If you are a student, there is no other choice you have but to trust them while they feed you absolute hogwash! They force students in believing gravity is the result of mass and to see that students comply with the unequivocal acceptance of the brainwashing, they subject students to various tests and examinations. In those test they determine the degree the students' have developed by brainwashing and the levels they test goes by employing examination standards. They never prove this concept that it is a fact that it is the mass of the Earth that pulls the body down while all formulae used in science prove otherwise by not using the mass of the Earth. While knowing this all too well Newtonians insist on students accepting Newton and still students have to acknowledge these concepts as truthful facts. If you think those in charge of astrophysics are the pillars of trust, then get wise by reading the following. What you are about to read is simply mystifyingly simple and yet to this day I have not had the privilege to challenge one academic any where that had the honesty to admit to the fact of Newton being wrong. After you have considered the following you might agree with me that even small children can reach a higher level of clear-minded logic and find more sensibility than what those scientists promoting astrophysics have because science lives in a make believe fool's paradise. One such an example is to put space travel and extra terrestrial life forward as even a remote scientific possibility. That is departing from any possible sane minded thinking. Even to consider space travel as an option shows the level of not even understanding the most basic principle behind gravity as we find the comet proves.

Only when and after proving that a student has totally lost all ability to think for him or her self may a student be promoted into the ranks of their sublime intellectual group. This form of accepting someone into their league, they gave the name as being a post graduate. The sifting process they named examinations. You write on paper what they tell you and never question their opinion and after passing that examination, only then will you ever enter their sphere of intellectual brotherhood. If this was not true, then how could all the misconception I show in this book remain on paper and be taught in Universities for all these centuries? Are there so many misconceptions as I claim there are. Does this sound far fetched? Then you better read on and I will remove your blindfold and show you what a world of deception the Academic Physicist force on us. Read the following and see how they, the high and the mighty, those who think they can replace God and those who think they can think on our behalf and to tell us what to think, read how much they are clowns and the jokers in society. Read how little are they, the Academic Physicists, able to understand concepts about Creation while they think they are able to replace God by using their superior intellect.

If you are a student in the science of physics, then ask your Educated Masters to please explain the following abnormalities you are about to read in this book and insist on a clear explanation about the inconsistencies they promote while tutoring physics as if the physics they present are the most flawless and accurate institution there has ever been. Ask those academics supporting Newton about the following flaws that no one, except me, ever mention. Get them to explain the inconsistencies they never talk about. Wise up and confront those charged with tutoring physics and see who should you believe. Then get informed instead of brainwashed.

Should anyone require more or better explaining, I would advise that person to purchase any of my books holding the title as an Open Letter On Gravity. A flying object is under this gravity control of movement and it is this that has crafts fly and cars requiring down force by the aid of aerodynamic devices.

As soon as the idea of gravity connects, with the idea vested in thought of mass by sharing the split concept in a unifying bout, then I say that gravity there is not. I say what is defined as gravity is not gravity and then what is acting as gravity, is not named correctly.

Gravity is defined as a force that is present in mass pulling mass and it is that entire idea that there is no evidence of. When I refer to gravity, everyone grabs on a cultural notion of a concept they formed and in that concept they link the smallest part of the concept to become and represent the overall gigantic principle and by knowing one line, everyone has the opinion that anyone then is the absolute master on the idea of gravity. When I freeze any substance, the substance contracts to a liquid and with more cooling it contracts to a frozen state of ice. The gas expanded more than what the solid did because the gas is hotter than the solid is. When we form the opinion that the outer space expanded to the limits, the idea springs to mind that outer space is freezing cold. When I say the Sun freezes hydrogen to a liquid because my eyes see the liquid squirting from the Sun, I am dangerously mentally impaired, since the Sun is blistering hot. Then through this culture my effort to say gravity is motion and motion is the cooling of overheating and being without motion or having too little motion results in overheating where such overheating leads to rapid and uncontrolled expansion and thus, the expanding Universe goes wasted. Everyone has the opinion that where gravity is the strongest such as the case is it is in the centre of the Sun or at the centre of the Earth, such a place is extremely hot and where gravity is least that place is unbearably cold.

Students tell your Professors to stop deceiving and stop trying to control your minds with their fraud. Those Academics tutoring you are telling you facts about gravity that has never been proven.
That is mind control.

They wish for you to accept facts on gravity that they hold as the truth. Should you question that mass produces gravity, they will expel you from University by letting you fail your examinations and it was never proven. Academics do put mind control to work on unsuspecting students by forcing students never to question the legality of statements they offer as being sound and correct.

For the first time ever and this statement takes time back as time runs further back than the time when and since Newton introduced gravity, there now is a logical and simple explanation as to what gravity is and why there is gravity. Before I achieved that discovery, I firstly had to find the centre of the Universe because it is there that I could locate gravity. I can now show how gravity forms because I have detected the centre of the Universe. Academics guarding physics will never allow an outsider to enter their domain without the intruder paying a heavy price and in this matter I was the intruder. I argue that if it is the correct practise to use $F = G \dfrac{M_1 M_2}{r^2}$ to calculate gravity, then the radius holding the gravitational constant must lead one to the centre of the Universe. If the Sun for instance has mass that is apart from the Earth and the Earth also has mass and there is a gravitational constant between the Sun's mass and the Earth's mass, we have the radius in that location. It then must be the gravitational constant that fills the space that the radius holds. Through my venture I discovered one person who knows what gravity is!

Ask yourself the following: If gravity pulls towards a centre and gravity holds the Universe attached, the question arising from that simplistic answer is then … where is the centre of the universe?

Science sees to it that Kepler stays the least appreciated Cosmologist where as in truth Kepler proved gravity, proved singularity, proved space-time, proved the Big Bang, proved every dynamic most of the wise persons afterwards thought about.

Yet, no one gave Kepler any recognition up to now because science denies Kepler his limelight.

I am sure every one knows there is a mistake in science but fails to recognise the fact. After almost three decades of researching I realised what the mistake is that hampers science. The fact that there is a mistake hampering physics must be obvious to everyone that ever had anything to with science or had any commitment with science or had any involvement in science. I tried to share my view with those in charge of physics. Being nice to those egomaniacs got me no where but frustrated. Because of that reason I am lately approaching them on the same mentality they have. If we are brush and honest, straight to the point and opinionated, then that goes both ways and we approach one another in equal manner and on equal terms. I realise any person reading this for the first time would be shocked by the manner that I hold, but believe me, with the high regard academics and especially those in the physics office have about their status, they believe no one dare approach their presence without bowing face down and crawling as one enters their presence. Academics have the opinion that only they can think… and to their disgust academics in charge of mainstream physics find little old me also thinking that I too, have the ability to think… and I think they are corrupt in their attitude and in the way they run their department. They hide the truth by brainwashing students!

I know my approach is inflammatory, rude, aggressive and distasteful but believe me, being nice and polite got me nowhere. Those dictating science has got such boosted egos that it is high time we start to discard diplomacy and get on with the truth. To them my having an honest attitude while I am trying to be polite to them and conveying what I have to say in diplomatic terms, to them only means being submissive. Seeing the nice way of discussing got me nowhere without progress to show and now it is time we (them and I) change or diplomatic attitude. The time has arrived where they receive the treatment they deserve and they come to realise they are not as superior as they hold their status of superiority, being self opinionated about their godliness while walking amongst us mere mortals and sharing air with fools such as the rest of us. I stopped calling Newton a mistake and after all the fighting that got me nowhere I now call Newton a fraud! If you wonder why I call this perceived as-if-to-be mathematical demi – god a fraud, then it simply is because that is precisely what he is. I call the one acting as a stand-in-for-god and seen from a human point as the-one-person-chosen-to-be-replacing-god as far as physics religion goes the one being just a super common trickster; this person I refer to is the one who is going around using the more common name of Sir Isaac Newton. I call every aspect that is super-humanising Newton's insight into physics treachery and deception committed on the highest level mankind can steep down too.

This approach is my way of coming to the point and fighting with no punching spared and taking gloves off. It is me fulfilling a promise I made two years ago, in September 2005. This is a promise I made to Newton physicists the last time they rejected my academic approach. After they rejected my explanation about the truth in physics I promised them I am coming back with a fight. I promised them (the nine of them, everyone heading a physics department at a University in South Africa with each filling a controlling position in presenting physics) a fight while I said I was going public and blowing their bubble of pretence. I promised them I am going to uncover their cave that is hiding the rotting bones they call physics.

In the beginning I tried a very subtle approach trying to handle those Academics tutoring in the subject of Physics with utmost admiration and uttermost respect while trying approaching them very gently in their esteemed positions and operating extremely cautiously while sensitively showing those I approach that there is a mistake in the approach science makes with science's understanding of the concept underwriting every possible principle in physics as a cosmic fact. I showed them that Newton had no idea what gravity is and neither did they ever have any vague clue about gravity or show the least ability in understanding cosmology, but I did it as gentle as I was able to. It didn't work! Then following that this to-the-point and cutting-no-corners approach came as the final straw broke the camels' back that was carrying my patience with their megalomania and there are hundreds of camels with broken backs this far.

Every time when I did my approaching and my effort to proselytise those in need of converting, the response I received was one of getting blasted back into the hole they saw where I crawled from. Their response was to let me know Newton was beyond the ability of doing anything wrong and Newton's perceived infallibility was the restraint that I walked into forming a wall of resistance which was called Isaac Newton. At first I tried to avoid Newton, but they kept throwing Isaac Newton's infallibility at me. Then I tried to get around Isaac Newton. I tried to get to go underneath Isaac Newton. I tried to leap frog Isaac Newton. Whatever I tried to do, it had me confronting Isaac Newton. It was a case of mad-Simon says or in this case Isaac Newton says and after saying Isaac Newton says all form of logic being present in the minds attending, then all sensibility present in the meeting then left the room. I am not converted to the religiosity of Isaac Newton by any academic qualification and found the devotion there was in forsaking all form of thinking in favour of Isaac Newton's religiosity being very absurd.

Then finally, after years of trying to persuade and later to convince academics about their unapproachable religiosity about the demi - god called the flawlessness of Isaac Newton, I realised I had to get my knife into the heart of the problem and if it then is killing the religiosity called Isaac Newton, then I shall do accordingly and I shall show the world what a fraud Isaac Newton is and what a lot of criminal thugs those backstabbers and common crooks are. I am referring to those Mafiosi professing their skills as experts in the subject of physics and what gangsters they truly are.

If you are a Newtonian academic in physics, please put the book down for I have had enough of your type of thuggery and criminal deception to last me one complete life time. I shall show students how they are brainwashed and being mind controlled into religiously believing in Isaac Newton while they are conditioned in accepting that Isaac Newton is beyond approach as he is flawless in every way and proven beyond doubt. Here comes the surprise that (I hope) will cure physics from its dose of fraud which became the disease of atheism that science has been suffering from for many centuries to the past. I wish to share with you how it came about that I got a breakthrough in my concept of understanding physics correctly. From my first encounter with physics I suspected that there was a gross error present in physics.

According to Newton's hypothesis, objects in space hold mass and that is what produces gravity. This suggested hypothesis (that is all it is) is completely unproven and a fallacy void from truth. There are three factors that is mathematically part of a three dimensional cosmos. Even if you are Isaac Newton, you can't remove one of the three just because you wish to con people into believing some nonsense story of mass bringing about gravity. All three are valid in relation to the formula. Kepler said $a^3 = k \, T^2$ and it is mathematical cheating to remove any one of the factors $a^3 = T^2$ **(k=0)** because then the table Kepler calculated is invalid…and believe me the table is valid and that makes Newton's arguments about $a^3 = T^2$ **(k=0)** very invalid. Remove one of the factors and the formula disappears.

PLANET	PERIOD (Years) (T)	MOVEMENT (T^2)	DISTANCE k	SPACE (a^3)	RATIO
Mercury	0.241	0.058	0.39	0.059	0.983
Venus	0.615	0.378	0.728	0.381	0.992
Earth	1.000	1.000	1.000	1.000	1.000
Mars	1.881	3.54	1.524	3.54	1.000
Jupiter	11.86	140.66	5.20	140.6	1.000
Saturn	29.46	867.9	9.54	868.25	0.999
Uranus	84.008	7069	19.19	7067	1.000
Neptune	164.8	27159	30.07	27189	0.999
Pluto	248.4	61703	39.46	61443	1.004

According to the chart Kepler produced we can clearly see that according to Kepler's finding, gravity is round about 9.86 and I prove that gravity is Π^2 which then is the same thing.

Kepler's finding on the other hand shows no proof of mass playing any part whatsoever in any factor concerning solar orbital objects. There is no indication of the presence of mass in any planet or cosmic object rotating the Sun. They are located at random with no mass favouring any one as to have any particular position. They all spin around the Sun holding very evenly matched velocity per time unit while travelling. There is the second smallest nearest to the Sun with the biggest somewhere in between and then the smallest on the very outside. They are scattered honouring no particular mass.

If mass was present and mass had an influence then either they had to arrange locations accordingly or spin faster and slower according to mass. There is no mass given to planets and academics in physics claiming the presence of mass are cheats and fraudsters. That is where the brainwashing comes about. Academics force students to accept what no one can ever prove in order to uphold Newton's fraudulent statements about mass that supposedly must form   gravity. The following

sketch supposedly proves that $\dfrac{dJ}{dt} = 0$. If there is one

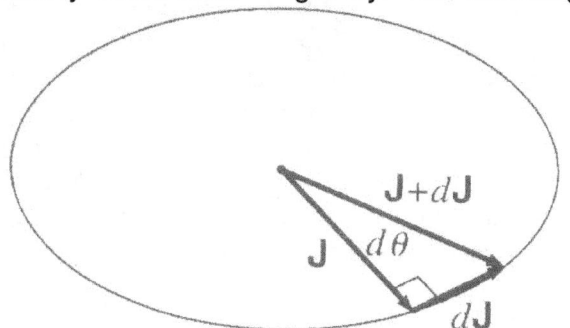

mathematician (not one person in physics compromised with being Newtonian) that can approve of this mathematical statement in any other sample but Newton's say so, then let such a person come forward!

Using some bizarre nonsense argument Newton claimed that through the spin of an object the radius

become irrelevant thus $\dfrac{dJ}{dt} = 0$. This is madness and I prove how insane and mathematically unstable thus argument is. Ask yourself how would one go about to divide one factor into another and then from such a dividing get a result of nothing. The argument Newton took can only apply the instant the object receives mass. Then the object loses individual motion and accept the motion of the Earth, thus forfeiting singularity by accepting the Earth's singularity and with that, accepting to have mass as a factor forming part of the earth's movement...but going into detail about this comment, I do in **Sir Newton: A Conspiracy to Defraud Science**. Then the argument changes to

$$\dfrac{dJ}{dt} = 1^0$$ in and only in that form it represents singularity.

But I explain that in the book mentioned as well as Newton's Fraud which can be ordered directly from my Web Page. He did this to allot mass left right and centre as it pleased his fancy and as he wished and the criteria he used was the visible size of the planets. He never took motion into consideration. No person can prove that planets have mass. If you are one of those thinking that just because your body or any other body puts claim to a space it holds in space and therefore space occupied grants the body, either yours or any other body, a certain mass, then think again. That is miles from the truth. Get into a high speed lift while you stand on a scale and see how your mass increases or decreases as the lift accelerates or decelerates...and don't come with the hoax that mass is not weight but something else. If it is space that forms mass the measure then used for mass should be in cubic whatever. They use weight to calculate mass and when those cheats don't make sense because they are painted into a corner they come with all sorts of crooked diversions that divert from the truth and try to swindle logic out of the argument just to put their deception into a shadow. This deception that mass is not weight but the measure of mass is the same as the weight, is all part of the academics in physic's criminal deception. They change their stories as it pleases them and never adhere to their own rules they lay down.

If movement can change mass, then the body is not married to a specific mass or weight just because the body occupies a specific volume of space. This too, is part of their trickery to make nonsense of science in order to favour their demy – god's corruption and betrayal of true science.

Let's quickly look at the following statement that $\dfrac{dJ}{dt} = 0$. There is no possibility that any person holding a sane mind can acknowledge that one can divide anything into any other thing and get zero. That is not possible and doing that is conning persons into believing hogwash. If there is one mathematician or physicist that can prove that this dividing of anything into another thing can possibly realise zero or nil as the answer, I would love to see that person do that! Please show me where this occur other than in Newton's imagination while he is dreaming under an apple tree.

 While being on Earth, there is a huge difference in mass between a wheel barrow and a steam locomotive. The wheel barrow is more hollow space than metal while the train is all metal covering very little emptiness. Yet, if and when the two objects fall freely we know the objects will travel at the same pace and land on Earth the very same instant on the condition that the two objects were dropped the very instant and travelled the very same distance and the second condition is that the two objects travelled the same distance. In that mass plays no part while they fall from the sky through the air. The emptiness that is filling the inside of the wheel barrow bucket falls equal with the metal forming the wheel barrow body.

I prove in my other books that it is not the object that falls, but it is the space the object holds that falls. The object holding space moves down with the space it holds in correlation with all other space surrounding the space the object holds and all the space moves down and not only the object in the space.

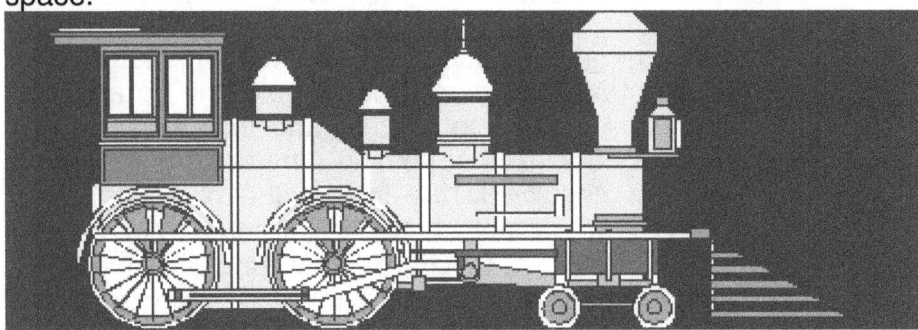 Then there is this nonsense that objects fall in accordance to mass because they say mass brings about gravity and gravity attracts objects. That is fraud and falsifying Galileo's work. Galileo said all things fall equal. I would love the see any of these crooks show how any object with a vast difference in mass can fall simultaneously and still fall according to the mass it has. It is the same situation we have with planets.

 There is no indication of the presence of mass in any part of the truth and because I have only Newton's' say so as proof, I have to accept that gravity is the consequence of mass. I want more than only Newton's word on the matter. That is rubbish spread by these fraudsters called Physicists. They are cheats, tricksters and con artists. If anything falls according to mass then all things must fall differently. If all things fall equal then mass can't have any role to play in the forming of gravity. It is as simple as that! Telling it the way Newton told it has criminal intentions of misleading others.

Years ago I was reading of a remark Einstein made about his realisation while he was a patent clerk. Einstein realised that had he (Einstein) fell from the window of the patent office, then he (Einstein) would feel as if he was as weightless as a chair and a pen falling alongside Einstein down the building.

Then I then realised Einstein felt weightless because he was falling and part of falling was feeling what was happening to him. He was not pretending to fall whereby he then would feel as if…he was really falling and with that there is no as ifs. What he experienced came by means of what he was experiencing. If Einstein was experiencing weightlessness, it would be because he was weightless while falling. Einstein would not imagine the weightless ness because Einstein was truly falling. He was at that moment truly weightless. Einstein, the pen, and the chair had the same weight since they were all weighing the same. All three items would be equally weightless during the falling…that was what Galileo found because objects of different size and different mass travel equal while descending. The bigger objects do not fall quicker than a smaller object and that can only be attributed to one fact; it can only be true if they weighed the same while falling.

From this, one can deduct that gravity is motion or the intent to commit motion and mass is when the motion of gravity is frustrated by blocking the continuing of the motion. Gravity is motion of space and mass is the restricting of the motion of space. Having mass does not bring about gravity but it does restrict gravity's motion. Gravity produces mass but mass does not produce gravity. Mass is the restraining motion and gravity is material moving about. Mass only comes into the application when two objects filled with space moves into a position where both want to claim space the other occupy. In essence it still is the frustration of motion and the commitment to move once the blocking of space is relinquished.

After reading this I then realised that gravity is not mass orientated, but gravity is motion differentiation between objects. While falling, the object moves less or slower in the direction that the Earth rotates and will fall in the direction of the Earth centre until such a time as the movement of the object is in synchronising with the speed that the Earth spins. If not, the object will land on the Earth surface at the edge of the Earth and that will bring about having mass. The gravity applies as speed that is putting time in relation to the distance travelled and distance travelled is space. While the object is in a process of falling, the motion confirms gravity, both by getting the object's distance on land in which the object travels in harmony with the Earth that conducts all the spinning taking place at that point. That will reduce the height in which the object spins until it lands on the Earth and then can't reduce such reducing of the travelling by landing any further. It has to do with specific density. If the specific density is increased by filling the object with helium we will find there arrives a point where the conducted speed is at a level that the Earth no longer will claim the body into having mass. When motion downward ends and the Earth disallows any further movement to secure a better specific density in relation to rotating movement, then mass sets in and becomes what is then the point holding gravity by virtue of mass where the constraining of the object takes place to secure frustration of further movement and the Earth's motion annexes the object's freedom. While experiencing mass the motion is still there but now incarcerated by mass and locked onto the Earth by the rotation of the Earth and the superior or equal specific density of the Earth. By connecting to the Earth, the motion that the object is experiencing is what nails the object to the Earth by the force of mass and the object is then experiencing mass and not falling further through the loss of downward movement and now only conducts with the Earth rotating side-on movement. In this the downward movement is not lost altogether but remains as latent movement that is detectable, where the movement is in the form that we experience as the body in gravity having a tendency to move although the object in mass is applying by forcing the downward motion to stand still. While the object is in mass and seems to be as if it is resting the tendency to move downward. In all, the body surges downwards while movement is still applicable but that tendency to continue to move downwards is the tendency he named mass.

However, mass then restricts motion and becomes motion tendency. While falling, gravity applies as equal motion to all objects relying to place all objects in relation to specific density and because of this motion counteracts any size, mass or weight by making everything able to fall equal in specific density. When falling, the object is either equal to what might be in the air according to allowed specific density, or has more than the specific minimum required density that is what is allowed to serve as the minimum required specific density and therefore will spiral down to the Earth. When the Earth restrains further downward motion of the object that comes as the result of finding an allocated position of motion according to the specific density of the falling object, this readjusting of allocated position is stopped from conducting further downward or readjusting movement and all such further movement of gravity is hindered in the form we call mass. The falling object remains individual and still tends to move while Earth individuality resists movement. Further movement is disallowed as other material fill space. While the bonding of the atoms forming the object will secure any further deforming the object will remain to be independent but it is this bonding that is the value of the specific density of the object applying. By securing a place on the Earth, the falling object will finally rest and from that motion resistance comes mass.

While falling, the object is experiencing gravity because the object is in gravity but when on the soil the object experience, mass which is the restricting of gravity or motion of the space filled with material.

Moreover, I came to another conclusion of equal importance. When any person is standing on any place anywhere, while viewing the Universe, that person is filling the centre of the Universe. Let's get more personal. When you, the person that is reading this, are standing at night and are looking at the Universe you are seeing the Universe from the centre of the Universe. All the light, every single beam that ever left any destiny at any time acknowledges this fact. You are the most important person in the Universe because you are holding the most important position in the Universe. All the light that comes across all of space runs directly in a straight line towards you filling the centre of the Universe. Not excluding the effort of one photon, all light is heading to meet you where you are in that centre spot and not one photon will pass you by. Not one photon dare miss you because if they do they miss the effort that all light has to accomplish and that is to locate you as the person filling the centre of the Universe. If you find this funny, or laughable you are in for a shock because this is what gravity is and this principle dictates gravity. It is the most complex issue one can imagine and expanding on this thought takes thousands of pages. It forms the crux of all cosmic principles and embraces every successful and meaningful theory ever used to explain the Universe. Without taking this aspect into account, there is no valid explanation available to understand the cosmos. Al the light coming from wherever meets the point you fill in time and in space. For all the light travelling, you hold the spot it was on route to.

Should you decide to shift your position to any other place in the Universe, you will shift the centre of the Universe to that location as well. If you install a camera on Mars, the light is obliged to acknowledge your relocating the centre of the Universe at your will to reposition you're being that centre of the Universe. All the light that ever left its destination crossing the vast spaces of the Universe, excluding no particular light, travelled all the way just to find you filling the centre of the Universe, right where you are. By you're standing anywhere, you fill the centre of the Universe, and the entire Universe admits to that because all the light comes to meet you there. If you shift from the North Pole to the South Pole you will shift the centre of the Universe because all the light travelling throughout the Universe will find you where you then moved the centre of the Universe! The light left its destination billion years ago as it travelled through space at the speed of light anxious to acknowledge you're being in the very centre of the Universe. No photon will pass you by where you are in the centre of the Universe. No wonder every person born has the idea they were born to fill the centre of the Universe, which we do fill. The Universe is spinning around you or me, which is filling a centre where all motion is connected. That is the Coanda effect on the utter-most grandest scale imaginable; nevertheless it is only a manifestation of the Coanda effect. It implicates gravity as wide as can be...

Then I reviewed the Universe. If gravity is motion, what causes motion? What stops motion? That answer is in the Black Hole. If a star is about fusing atoms thereby growing, what happens when all the atoms fused into one all-collective atom? What is the gravity if the star has one all-inclusive atom providing all the gravity that the star had when the star still had massive volumetric space? If all that space that once filled an entire giant star fused into one enormous gravity applying atom and that enormous force has been secured in the space that one atom holds, the atom would then show a force that would pull the surrounding Universe flat. Where does the gravity of the star end when all the atoms in the star became one giant atom? Gravity is smallest where space is least. Where space of an entire massive star is left in the size of one atom, the gravity coming from that will pull the Universe flat at that point.

Coming to the conclusion about gravity being motion and mass being the restriction of motion was the easy part. What produced the motion and what prevented the restriction from overcoming the motion was the tough part. Figuring out why everything was on the move and where did the motion stop, that was the part that took some figuring and some explaining. What made gravity move and why does gravity move…the answers are in the four phenomena never yet explained to satisfaction but now turns out to be the cradle of gravity.

Gravity is The Roche limit,
    Gravity is The Lagrangian system
       Gravity is The Titius Bode law
          Gravity is The Coanda affect

Gravity as the Roche limit forms the principle that is responsible for producing the phenomenon named as the sound barrier. Read the book and find out why this is the case. The Titius Bode law, the Lagrangian positioning and the Roche limit phenomena all combine in one unit forming the Coanda effect principle. That constitutes in gravity. However, explaining this in specific detail requires books of pictures and space and that is not possible in this printing format.

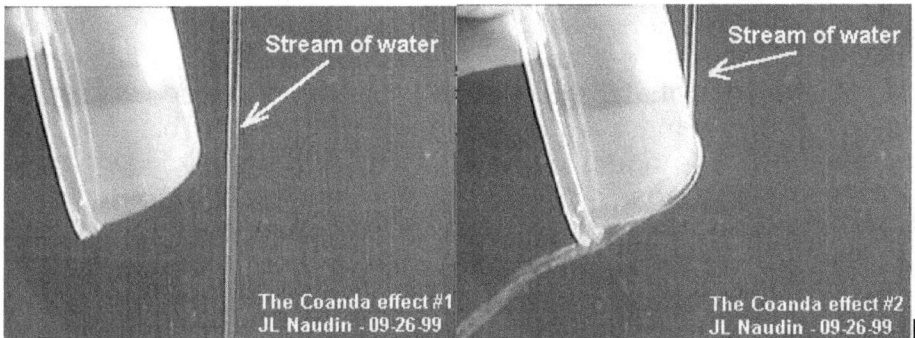

Newton's claims about the principles that he declared is responsible for guiding physics carries no validated proof and only after I realised that, was I able to start forming another line of thought on gravity. This had the purpose of confronting the corner stone of modern physics and at first I tried desperately to do just that. At first I was not confrontational towards Academics in physics and avoided any indication about disagreeing with Newton, although avoiding showing my disagreements was also totally impossible too, but every time I approached academics with my new concept the academics always threw Newton at me. Facing Newton or facing defeat became a two-sided blade and I had to start to confront them by confronting Newton, with which I was in disagreement from the start. At first I was reluctant to voice any opinion about the god-sent to physics or consider the matter of how far I was prepared to challenge Newton, because Newton was and is an icon. But slowly it dawned on me that if I had any serious plans to introduce my ideas I had to dispute Newton's gravity principles and kill the monster by removing the head. I began to dispute Newton slightly but very openly and…it did not bring results. When the slight confrontation did not bring results I finally decided to go all the way and show the inconsistencies that were prevailing in Newtonian science. That did not work either and it brought me the same results as before whereby I decided to go public and straight to John and Jane Dow and avoid the arrogant academics with only one motto they serve and that is their autocracy and in particular their megalomania especially to my case as well as me in person. I wrote them (nine in total) letters in which I warned them that I was going public to show the extent of their dishonesty in their Newtonian approach and lack of substance and proof their physics have. The lack of honesty and furthermore the absolute dishonest on their part is there whether I avoid it or attack it; the inconsistencies are part of forming the basis for modern accepted science.

This process I now described is explained in a paragraph or less and it seems I got that far in a breath or two, but getting this far took me the best part of seven years to get to, I tried my best not to attack them or Newton, but I was left with the option to leave the project and lose thirty years of work and then fail after I concluded an answer on every aspect they never even thought of or take them on and dish out what they should have received years ago, made me decide on the latter. After being avoided and taunted by their powerful positions and arrogance vested in their mentality they show in regard to their positions as well as the disregard they show in the mentality towards others I slowly concluded that only and after I can get the general public to see what they hide, will I get a response from the Mater's of fraud. First I had to show the general public the true colours of the academics in physics and get every one to see how incorrect Newton is, and only then do I stand any chance to introduce my line of thought. I am so sure of the ideas that I propose of being correct that I dare any one to disprove any part or the entirety that my concepts about cosmology form! But that can only come about when I can get an audience to see how I expose Newton for what Newton was and it is in that where I find no luck. I can't find one academic with influence who is brave enough to stand up and face my attack on Newton and argue me down or prove me wrong in a sound debate. Now I see frowning coming from everywhere because it is madness on my part to think the world is wrong and only I am correct!

I realise that it shows signs of madness on my part and in my thinking to even regard any possibility that I am the only person on Earth that is correct and all others that ever studied physics are wrong, but mad as it seems, if that is what I have to say to find an audience to listen and to judge my case, then that is what I say. I don't say this lightly or without understanding the enormity of what I suggest is going on, but be that as it may seem, it is the truth without question that Newton went on for three hundred and fifty years defrauding science with no one testing his claims. Argue me down or prove me wrong but don't discount me before hearing me out and only after considerable consideration while studying my arguments, then form an opinion that disputes what I say but when disputing what I say, do it while confronting me in a sound argument when proving me incorrect! This, not one academic could achieve and I challenge the lot to do so. But do it after studying all my work and being in a position to account for all the details I propose. Don't just dismiss me because I dismiss Newton, because following that road is the way of the coward and the mentally impaired. Read my challenge about the correctness of Newton's proposals when he brought no more than suggestions into science and when I dispute Newton, then take me on by proving Newton correct... do it just once... prove Newton correct just once...prove that his formula is working and that his principles apply on the grounds he principled his ideas.

I was able to detect Newton's judgement errors by not being absorbed by the age old culture that drives science. I went about and tried to prove Newton and when that was not happening I tried to apply Newton's ideas into the greater fields of cosmology. That also wasn't possible. I tried to amalgamate the four cosmic principles applying in cosmology with what Newton said was happening in the cosmos with mass and with gravity and in light of what the cosmos showed was happening Newton just wasn't happening! Notwithstanding the pose Mainstream physics try to uphold, the entirety of physics still use the idea of magical forces intervening in nature and they still base concepts on unexplained novelties. Think of finding four unexplained forces going around and influencing persons in an unexplainable manner except that the magic of gravity keeps people attracted to the Earth. To say the least, the concepts physics use in terms of Newton would not even be acceptable to children in the modern informed era we live in, I challenge any person to prove Newton, not to accept Newton but to undoubtedly prove Newton correct! Prove how Newton's formula of mass forming the force of gravity can apply as Newton said it does! I recognised the impossible double standards Mainstream physics apply to promote their much shady explaining. In short I tested Newton's principles and found the principles to be wanting.

The inconsistencies Newton introduced brought science double vision and to compensate for these bogus truths supporting their incredible theories, they simplify issues to such a level where what they embark on, is the meaningless acceptance of the unproven and they proclaim to understand what are meaningless inconsistencies and to achieve this they create scenarios which use the entanglement of deception. Prove the attraction Newton said was enforcing gravity that is pulling by mass and is gathering planets by contracting the diameter between planets. Show how much the Moon came closer to the Earth since the time of Kepler. Show proven distances taken by radar tracking and indicate just how accurate Newton was. Show how much the Moon came closer to the Earth since the time of the Moon walk in 1969. The figures are available but are kept in a grave of silence where no one ever speaks about what science found applies and how much the distance between the Earth and the Moon is shrinking as Newton said is happening or then how much is the is expanding which will contradict the very principles Newton brought about! What they declare as unwavering facts can't even be supported in some form when tested by a silly test as to show that the distance between the Earth and the Moon is shrinking. Even the least degree of verification of correctness is absent when trying to find support of Newton and Newton lacks all evidence of authentication in any investigation of even the simplest terms. It is as if they never read with interest that which they explain when they embark on explaining Newton and they never scrutinise that which they advocate when they teach Newton's principles applying. They give values that are senseless and the very values they use make that which they say meaningless.

In this book I am going to investigate how much truth there is in mass pulling by the force of gravity. To most if not to all of the persons reading this, such a venture of investigating Newton is time wasted and just the thought about me embarking on the investigation of the issue is totally senseless to investigate. It is senseless because the concept it carries became accepted as household practise and life science from where it proceeded to become everyday culture in every person's mind. The worst part is that the group of people normally considered as the wisest bunch there is, never did prudent testing on Newtonian presumptions, while to test the presumptions is most easy to do. I will not believe that academics who live up to the veneer of being the best mathematical intellectuals on Earth, never though of testing Newton's very simple formula and in that disregard the formula because of the incorrectness the formula holds.

One should think that a department such as physics is run by level headed persons, persons who would distance their field of focus from magical forces. Yet, when I showed them that science did not progress much since the days of Newton who lived in a time when reading your horoscope was done by the most intellectuals that life had to offer, and magical healing was administrated by the most-educated in society and what science understood about chemistry placed science at the time little apart from sorcery. Then by my accusing them of being backwards, it is me that they discard as being incoherent in my opinion forming. Those same academics who shout as much blame at the Pope for ridiculing Galileo are the same as those at present who are denying Galileo his credit. They deny Galileo his rightful place in science every time they portray the views of Galileo as being equal to what Newton believed. Galileo said things fall equal notwithstanding size and Newton said all things fall in accordance with mass and only the brainwashed mind of the "sober thinking" Newtonian see it as the two having the same opinion. They shout at the Church and demand that the Pope should apologise for the behaviour of the Church against a person such as Galileo when he was persecuted for having an opinion that contradicted the popular view at the time, but in my case I am just as much opposing the popular view of the modern era which is the view they all abide by, yet their attitude towards me is even more vindictive than what the Church was when the Church went against Galileo. I knew from the start that I would find much resentment from certain quarters, but resentment is not even a concept compared to the treatment I received from all quarters this far. Yet, not once did one person show scientific evidence of what is forming the working principles of the four forces they say is at work in science. Never is there only one who can indicate how the one mass will find a manner to pull across a vast distance formed by (as they say it is) nothing and still find a way to pull on another mass it has no physical contact with. Is that what one should expect from sober, level headed, to-the-point persons that only deal with truth and are only able to take into regard substantiated and proven facts? You decide if those academics claiming to be trustworthy on all levels are as sober in thought and conduct, as they would like us all to believe they are?

The manner of regard to life that the Academic Physicist holds and the outlook on life that the followers of Newton physics have (I call them plainly Newtonians and to me they are sheepish because they resemble to the image that to me seems the same as sheep running after their leader without having the ability to think for one second any thought spawned out of personal intellect) is quite the opposite of what I think of them. They keep their forming the establishment of the order the Academic Physicist in high regard and consider their order to be the top thinkers in society. This religion that they practise of self promotion and sublimely self regarding their status being next to God has them so high that we down on Earth forming the waste of human garbage can be told anything and we will believe what they say just because they with their supreme intellect tell us to think what they wish us to think. This they do because we human waste living way down below their supremacy have not the ability to think and therefore they must think on our behalf. In their view and so far very correctly judged on their part, they, the persons being in the group that forms the Academic Physicists, believe very correctly that can dish up whatever they wish and we, those forming the group in the gutter, those that are mindless in their eyes, we will have to accept what they say without being allowed to form an opinion other than having the opinion they give us to have because in their view we are unable to have a mind other than what they are able to control. This attitude they have is the result of a relationship that worked for so long and the fact that it worked that long is what confirmed their opinion that we, the public, are fools to believe anything and everything because of blind stupidity.

But in spite of their aggravating conduct and mischief towards us, it is not because of a lack of insight and inability of controlling a mind that we have our childlike belief and blind trust in their opinions and which there was. It is the faith we have shown that they misused for their scandalous cheating. Our faith is what we have shown towards them and is that, which became used as the reason why we accepted what they said blindly. We didn't accept their word on the grounds of us being utterly stupid as they perceive us to be but our trust depended on our good nature and believing in their trustworthiness. This trust we have is brought on by a culture of trusting the King to do the people well and somewhere in every person's cultural past there was Kings who did us well in leadership. But their underestimating of our abilities is the testimony of their poor understanding and their weak insight ability which results from their arrogance and stupidity. You are about to see just how stupid they really are in the thinking aspect of science. It will become clear as you page along while reading! They didn't fool us half as much as they fooled themselves and you are about to read all about it. The fact that they could fool us for centuries didn't run on their intelligence being so much superior but served their purpose as it stemmed from the trust we had in them resulting from good intentions on our part. This betraying on their part and misusing the public's good nature to be used in schemes to get the public conned must end and I pray that this book form the first step in resisting the arrogance of the Academic Physicist.

Any one who is not forming any part of their group of the Academic Physicist holding degrees to prove that the person is already part of the academic superior elite, those members in the group regard the rest of us who are socially outside this group to be part of the lowest order of mindless being and when any one has the desire to become part of their order and members of those that have minds with an ability to think, which is what only their group is capable of, such students have to accept what their academic superiors say when they say whatever they wish to say without having to prove the correctness of what should back their academic saying so and as a result of this students may never question what the academics say is the truth. The academics have a formula by which a person enters the ranks of the super-intellectual and this order they named examinations. By testing your ability to repeat their brainwashing without showing sense of doubt about flaws that may be present in the system of physics, such student may then be promoted into the ranks of the accepted graduates. Those that unconditionally accept Newton is the group the Brainy Bunch accepts as the future graduates. They are those intellectuals that form the next order that in future will try to promote Newton. Those that unconditionally accept Newton is the chosen ones gifted to become the sublime intellectual group that will rise to form the future and then become the Masters selected by those already forming the Brainy Bunch group. This promotion into their sublime ranks can only come after showing true understanding of Newton, which comes down to following Newton like an arse, never to question Newton's claims because they say accepting Newton equals understanding Newton and understanding Newton equals being sublimely intellectually gifted. What I found to be true is that accepting Newton is actually to be mindless and without a personal opinion. The sifting process which is introduced to see how much total obedience they have in their control over your mind of those students being tested by examination is a process that they named examinations. Examination is as follows: You repeat what they say without asking them ever to supply proof that would support Newton's gravity of attraction and when doing so, then by their examination while repeating the information that they supply and that you must repeat. Then with giving such replica of what they said must be done by never questioning their information or the standard they set down as a rule. Only after they are completely satisfied that all physics students have become mindless zombies and is in a state of trans, then only can those students who show obedient acceptance of Newton without ever questioning the validity of such accepting fall into the category of the accepted few. By showing total idiotic devotion will those in charge of the examinations allow that the pre-computed victims they call graduates pass the examination and after which such computed zombies will then be permitted to enter their sphere of the intellectual brotherhood. If you don't believe me, sit back and think what will happen if you don't repeat what they say you have to believe about Newton. Ask them to prove Newton by explaining what brings on such attraction over vast distances they filled with nothing. Think what they will do to you in class when you insist on them proving Newton and you refuse to commit to what they say without them supplying the required proof. If you think back and find what I say is far fetched, then try it in class and see what happens to you. Insist on them showing what the so called graviton is and be sure to ask them as to where would you have to look if you wish to see this fairy tale sub-atomic particle they devised to prove gravity. See whether they will prove Newton or whether they would rather opt to chuck you out of their class.

Students in physics are conditioned to believe that mathematically one would go about to use Newton's formula $F = G\dfrac{M_1 M_2}{r^2}$ to calculate the force of gravity. This dogma is told by generations as if there was never doubt considering the universal proof that was brought the fact and the proof convinced all into accepting the formula as correct even beyond God. Put in the Earth's mass in place where it belongs and put in your mass in place where it should be and then divide that with the distance between your soles and the Earth measured in micro millimetres by the square thereof! Replace the symbols with figures and use the formula constructively as to prove its accepted accuracy. With figures replacing the symbols the formula is useless and is shambles.

In the book named an ***Open Letter on Gravity Part 1 and Part 2,*** I bring the solution to the mystery behind gravity. I tried in vain to introduce the principles I find valid to the academics in charge of astrophysics. Facts that Science present as being the uttermost explicit and unwavering truth, fail to bring any logic answers to so many questions that it should address. It fails to have substance in addressing the most basic and simple questions about gravity and physics. Yet, to every question science can't answer my approach does bring many solutions. The presentation and the delivery of the answers that I reach are understandable and simple where it serves both logical science and the truth. Since my answers do not match Newton and his misconception about gravity and that mass generates gravity, those in charge of science don't even bother to read my work. With their affixation on the corruption they portray, I can do little to the giants where they are in the mighty positions they have and just because of that they can go about to sideline and ignore my work and this is notwithstanding the correctness that my work delivers compared to the utter failing that Newton's work shows. When confronted with my evidence and they have to match my work with the hypocrisy and misleading nature of Newtonian cosmology, their defence in substantiating their claims is to ignore me. Since I do not applaud mainstream science and the clear fraud they embrace and fraud it is that they embrace, I am silenced. Why is it that my work is going unrecognised or even in the least goes never debated and never commented on...it is because it will then trash every article anyone has ever written about astrophysics and cosmology. They show little integrity when academics with such supposed high standing or then such as they should have, play a dishonest game where those in commanding positions will rather protect fraud and save their skins. They would rather protect the corruption they have than seek the truth and find honesty in physics. Those academics in charge would much rather protect their un-defendable ethos they maintain as forming the backbone in science and what gives their personal position legality although it is corrupt, than admit to the truth they find when they begin reading my work and in agreement they then have to back the truth my work brings. Doing that (accepting the truth in my work) will trash all work in cosmology delivered thus far and condemn it to the waste paper basket and render all work invalid and void. It will put the

Newtonian's bias and fraud into the place where it belongs. Considering that such acting will lose them money, those academics in controlling positions then will rather rape the truth in order to benefit from continuing to corrupt the student's minds further. If they wish to justify their inconstancies they have to attack my work and disprove the accuracy of my work. That they can't do. They then ignore my work because they can't attack my work. In that sense they also place their work beyond my approach, as they can simply ignore me as if I represent the plague while they carry on with little consequence to bother them. I challenge them to prove Newton correct and not just declare Newton being beyond reproach after all has seen the evidence I bring. After reading this all students must challenge them to defend what they can't or get honest.

$$F = G \frac{M_1 M_2}{r^2} \qquad\qquad F = \frac{r^2}{M_1 M_2}$$ This is the basis that Mainstream

science uses as the foundation of all physics anywhere. If this is wrong, then everything they have got to work with goes out the window. They put mass and the distance that parts objects in a relevancy, in other words the one is a ratio to the other. The one factor brings a measure to the other factor's value. The one cannot be without the other. The increase in one becomes the reducing of the other and the other way round also applies. When the distance is large, the influence of mass will be small and when the distance is small, the influence of mass will be overwhelming. Then they state we are in a Big Bang expanding of the entirety. Why then, when considering that if it is mass that produces an inclining force of contraction as Newton says there is going on, with all facts considered, why didn't the expanding stop before it started when the Universe was small. The gravity that was present in the first instant of time applying as the cosmos was born, had to be so enormous as never could be repeated afterwards because the radius was as small as never afterwards presented. After the first instant of birth the radius grew bigger and that reduces the influence that mass could have,

that is if Newton's formula $F = G \frac{M_1 M_2}{r^2}$ ever applied. Today using hindsight after the fact of

the exploding Universe became apparent by the studies Hubble brought to light did the lot of everything that is not implode as Newton would have us believe whereas, instead it did expand just as Hubble proved. The radius at the time of the first instant back then was no factor, which makes the gravity at the time a totality of unrivalled force. The radius being that insignificant leaves the mass unchallenged in asserting power in relation to the non-existing radius it had.

I dare any physicist to show me where they apply Newton's formula just and exactly as Sir Isaac Newton suggested gravity applies. Show me just once where the mass of the Earth is multiplied with

the mass of the object in normal physics. Show me just once how $F = \frac{r^2}{M_1 M_2}$ or $F \propto \frac{M_1 M_2}{r^2}$ where

one M represents the mass of the Earth while the other M represents the mass of the object and in this formula the end result will have a value of 9.81 Nm/s^2 … show just one example… where the use of the mass of the Earth comes into play. If multiplying the mass of the Earth with the mass of an object and dividing that with the distance parting the two mass factors does not deliver 9.81 Nm/s2,

then any claim by Newton indicating that $F \propto \frac{M_1 M_2}{r^2}$ is equal to gravity, such claiming constitutes to

deliberate fraud…even if Sir Isaac Newton said this. Prove that the mass of the Earth with the mass of an object and dividing that with the distance parting the two mass factors delivers 9.81 Nm/s^2 or admit physics is conducting fraud to protect Newton!

After reading this part my wife forewarned me that this chapter is monotonous…and yes, it would be because it is about my life all the while I am trying to get recognition about my work and the frustration that I have in everyone ignoring me because of a chap with the name Newton that could never do wrong…or so it is thought all along by everyone until I got an opinion contrary to that. Should you think this next chapter is monotonous and never ending, please think how I feel because it is vaguely detailing an account of the treatment I receive from the Physics academics? If you think you get a feeling of frustration while reading it, then it is simply my frustration boiling over and spilling on you. I have been advised by my wife to remove parts and so I did but I couldn't find reasons to remove the entire chapter just because it seems laborious. My quest has been on going for eight years non stop and should you wish to know what I am experiencing with the academics I fight, your reading time that this chapter will take and the part it would represent in relation to eight years of struggling with blind academics then should only be limited to a few hours that it would take you to read it, while it should take eight years of constantly finding utmost frustration for that is as long as I have been struggling to get my message across. I repeat that if you feel frustrated by reading along, then please consider that you are merely sharing the frustration I have been exposed to for the past eight years.

You may wonder why I would put a chapter such as this chapter is in the book in the first place if it only represents a feeling of shared agony and misery. The reason is to show those sanctimonious white washed rotten graves filling the posts of academics in physics their true nature. I wish to show that those hypocrites that shamelessly condemn the Roman Catholic Church for their misconduct against science in the case of Galileo Galilee are no better when they show even less tolerance when someone comes and disagrees with their pretender god Isaac Newton. If you get a feeling of getting dispirited, then please think how I feel and while you may experience it for a short while, while reading this chapter, mine is ongoing…

Circumstances beyond my control forced me to seek the help. I am challenging an adversary I can fight but I can never fight the opponent I chose as an opponent with the purpose to win by personal strength. All of the entire human race think my adversary is blameless and lily white, while I am the only one who knows better. I now require the force of the students and the logic left in their intellect to help me fight those most powerful. Those I challenge are much stronger than me. I have the best story and the biggest breakthrough seen in almost four hundred years in science and I can't get a single person to listen to what I say because they use their power to block me from telling the truth. They prevent my efforts in allowing the truth to be told. I have tried so many avenues in the past to get out of this predicament in which I find myself and have never been successful in conveying my discovery to other persons outside my immediate family. I have not even had the least of fortune to get my discovery announced and therefore I am very sceptic about having success this time, even with all the detailed planning I did.

I also am realistic enough to realise how this whole enterprise must seem to onlookers. If someone came to me and said he saw a man from the planet Mars fly a hovercraft over my house I would be extremely sceptical. If the man claims he is the only person in the entire world that has the ability to see such a craft flying, I'd say he was quite mad. If he keeps on shouting and claiming he is the only one who can see this event taking place on an hourly basis, then I'd be the first that would vote to have this person locked up in an asylum with many other loony viewers of such crafts. However, in my case I am really the only one that can see Newton being flawed and there is not one other person ever who were able to see this, except me. I realise that makes me far worse than the Mars viewer I created as an example to show that I realise how outrageous it must seem to the better informed than me and that I know about what is going on. There is no asylum with others such as me locked in it for public protection and that even thought of such a possibility of Isaac Newton being incorrect puts me in a ratio of one human born in three hundred and fifty years. Amongst all those who are recognised as the most intellectuals on Earth, going back three hundred and fifty years, according to my opinion, they all are wrong and only a stupid arse such as I have the ability to see what no one else were ever able to see, about Sir Isaac Newton's incorrect behaviour. I realise I may seem as a joke and the village fool where I have the audacity to claim sanity amongst great madness when I am the only one ever to come along and not only say Sir Isaac Newton is wrong. What I say (and I do say and claim this notwithstanding that I too see why others see in me symptoms valuating that others may think of me holding utter madness and showing signs of delusional thought) I still maintain that every person who ever studied physics which include the best minds ever born from a woman, they all are wrong. The world is marching to the tune of Sir Isaac Newton and everyone is in step but me, and I turn around and say every one is out of step but me, making me the only person on Earth being in step with every other person being out of step going back three and a half century. Look, I am not so soft in the head that I do not realise how this seems to those intellectuals who must read my work, but for God sake, stop viewing Newton as a god and just read what I say about the man. The man is wrong!

The story that I am about to convey has such magnitude that no one takes me serious although what I have to offer goes beyond any scientific breakthrough ever accomplished by any person at any point during the past history of human intellectual evolution ever since civilisation began. But I was handicapped by circumstances and influences far beyond my control, when every time I tried to convey my discovery to the public and because of enduring so many disappointments I have lived through, I am not very optimistic that I will find success going down the alley by trying (yet again) to seek for a publisher even though what I am now doing at this moment I had planned with so much care to detail and also with so much detail that from this intended contact with a publisher I wrote five books during all the planning that lead to this event.

With having so much success on the academic discovery level in uncovering what no person yet could accomplish, I also admit that I also endured so much failure in getting my work out and printed in the domain of the public that now I am a walking statement of frustration living in a world of rage. If I come across as being emotional, then it is because I am emotional due to my frustration which brought on my total resentment of academics in physics and the corruption they embark on while they try to hide behind a veneer of self-righteousness and when having this attitude of being holier than thou while what they protect is rotting flesh and stinking bones which they keep in a lily white coffin hidden at the back of a very dark cave. I have been frustrated by an adversary millions of times more powerful than me and I have no choice but to keep fighting because I have far too much to lose when I stop fighting, although I am the only one who are able to see what I see.

Eight years ago, in 1999, I realised what the discrepancy is in science and started to work on the solution. I saw how Newton went wrong and how to correct Newton's incorrect view in explaining gravity. Since that time I realised that just knowing the answers alone got me no where but it got me rejected and that rejection got me frustrated to such a point where I now feel that I am boiling over. If you think my aggressive behaviour is out of place, then please consider that fact that in the beginning I was not as venomous as I am now and I admit that I am now filled with poison to kill a yellow belly snake when thinking of my feelings I have developed towards all those in charge of physics. I wrote my first book on the subject but found no audience in Afrikaans and was advised by publishers I then approached with the Afrikaans manuscript to redo the project in English which I subsequently did in 2000. Writing the first manuscript in Afrikaans took me about a year. Redoing the manuscript in English plus developing it to be more informative took me all of the time the year 2000 presented me. After again approaching Publishers in South Africa, Van Schaiks, which is a well known and well respected South African Publisher, advised me to first seek Academic approval concerning my theory before any publisher would find motivation to go into any contract with me in such a publishing venture. Then I started to approach academics in the field of physics. That is where the problem started. Up to date I have contacted far more than one thousand six hundred all over the World.

While contacting that many, I have been running into an impenetrable wall the past eight years trying to fight in the most one sided fight you could ever imagine. I have discovered what gravity is and I found it is everything but what Newton claimed it is. I am able to mathematically prove my case and I also do it by means of employing true physics with which I prove what I discovered where such proof extends above and beyond any doubt. When saying that, I must also declare that at the same time I have come across a wall of academic resistance I can't even scratch the surface of. I can't dent the paint of this impenetrable wall they put up to surround Sir Isaac Newton with invincibility. One should think everyone is in festivity about this achievement I claim to have produced! Instead it is a brick wall I am running into as the Academic world of physics upholds their protection of the godly image of Sir Isaac Newton's uncompromisingly correctness. What I discovered gravity is, clashes head on with what Sir Isaac Newton supposed gravity is!

I dispute Newton and so do all students learning physics dispute Newton because Newton's arguments are an onslaught on ordinary unspoiled human intellect. Think of the resentment that students have towards Newton under normal conditions when they have to cope with understanding the Newton principles Mainstream science says are applying and how that confusion of what is possible and what Newton suggests is possible clashes with their intellect which makes them feel stupid. Students hate Newton because they don't understand Newton and for that they then subsequently are accused of not having the intellectual capacity to follow Newton. I am sure that every student coming from the past, going into the present and even including those forming a future generation of students will purchase a book that is showing that Newton's legitimacy is cracking up when exposed to some vivid scrutiny. This fact gives the book a selling potential like no other book in the past could do. Yet I am unable to find a publisher because publishers need academics to assure the correctness of the information in the book and academics would cover up Newton's errors at all costs.

Please bear with me, for if you do your reward will be that you will get wise and see Newton in a very different light, except if you are one of those performing the brainwashing academics unleash on unsuspecting students. What you are about to read can only be supported when students are subjected to massive brainwashing! In short I will now explain what I explain throughout the book and introduce you to information which you are about to receive for the very first time (if you are not a practising academic tutor in physics already) and is the reason why I selected a name such as **_Newton's Fraud_** or whatever name it will be named as. The Newtonian formula

$$F = G \frac{M_1 M_2}{r^2}$$ is the formula used by science to explain and define gravity. It says the that the

(**M₁ × M₂**) mass of one object pulls the mass of another object and this process is done in relation with a gravitational constant (**G**) (a supposed force keeping the Universe attached) and the pulling subsequently destroys the radius (**r²**) being between the objects finding attraction by the measure of mass. That says that objects **ALWAYS MOVE CLOSER _BY FORCE_** in relation to **_MASS_**. Newton submitted the suggestion that objects fall as MASS provides the force that will cause the falling by the inducing of a force he named gravity which he subsequently only proposed was the acting supositious force. I disprove this formula in so many ways in this book and I show that this formula and the ideas Newton introduced just don't stand up to even the smallest tests. Then, if Newton's idea on gravity has validity and mass is responsible for objects falling, then all objects that are in a process of falling must be subjected to mass and in that idea rests differentiation and discrimination in size and compactness producing speed variations. If any and all falling of bodies through the air going towards the Earth goes according to mass that creates a gravity force, then the result must be that the outcome is subject to the variation mass introduces and the influences coming about is the result of mass interfering in changing the gravity force being generated, this then must bring different speeds to cause substantial variation in the falling of different objects which are holding all the various different mass factors.

There can't be uniformity and conformity in the falling process of all objects. The conformity is not possible while such falling is the result of the discrepancy that mass has to inflict when enforcing gravity due to variations that must result by the measure in mass differentiations. This is a vital issue that science eludes and avoids addressing and the approach they take on this issue is one that carries the mark of deception. When tutoring this aspect in science that must indicate a dramatic difference between Galileo and Newton's approach on this issue they have all clever ways to avoid direct questioning by clouding the evidence by using as an example a feather that floats in air in relation to a hammer that falls directly. Confronting this particular issue is avoided in this way when science just runs around and never addresses the apparent differences that arise when it is scrutinised and in creating this deception they comfortably avoid confronting the issue.

This avoidance of confronting the issue which, when correctly addressed as I address it later on, will disprove the validity of Newton where I circumvent the manner of portraying the issue such as it was done during many centuries with such cunning as you will not believe. The way that it is presented in terms of a feather and a heavy hammer is indicative of deliberate fraud. The fact that objects fall due to conformity in the falling, science accepts but portrays a picture of deceit that mass brings a falling distinction and therefore equal falling doesn't happen under normal conditions, while they at the same time admit to Galileo's presentation that falling of all objects are equal in tempo, irrespective of size or any form of differentiation. While they promote the obscurity that Newton and Galileo is in harmony, the truth about their deceit is witnessed in the fact that the two can never have the same outlook on this issue. That I prove is a fact and also I show how big a part this is in the overall covering up of Newton's initial fraud. For that I am ignored whenever I write to Universities to state my case I present on gravity.

I have written several books in which I challenge the thought process of Mainstream physics and especially Sir Isaac Newton's arguments about physics. I am of the opinion that even though everyone thinks of Sir Isaac Newton in terms of the genius who established every aspect used in modern physics today, but in spite of every other person hailing Newton, I remain of the opinion that the man did not have a foggy clue about any of the principles driving the concept that he named as gravity, or what brought about gravity according to his explaining of what forms gravity. I am able to explain gravity but it doesn't even vaguely resemble Newton's version of gravity. I can explain gravity by proving my explanation with the use of simple mathematics. I use Johannes Kepler's formula to back up my statements. By using Johannes Kepler's formula I found a way to prove there are four phenomena found in the cosmos. There are the four phenomena applying in tandem that together forms gravity.
They are:
The Titius Bode law;
The Roche limi ;
The Lagrangian Point System and
The last as well as the combing of the four is presented as the Coanda effect.

As the phenomena don't support Newton's vision on cosmology, the phenomena has no support amongst Mainstream science although they did apply many with a positive results where the Titius Bode law was used in locating the missing planets at the time of their discovery. When they located unknown and undetected planets in the past, the existing of the phenomena was never disputed but when the argument of proving them comes to mind, and the fact that they denounce any viability Master Newton should receive, then they are dismissed as some coincidental abnormality occurring. But since it holds no similarity to Newton's view on science, Mainstream science rather disclaimed the validity of the phenomena than they would find fault with Newton's ideas. In the mind of science the cosmos can be wrong and God can be wrong but Newton can never be wrong.

In using the four correct principles correctly, which I back up with the correct mathematical interpretation thereof in support of the function that each phenomenon has in forming gravity, I did a far better job than what Sir Isaac Newton did and what I achieved is of a far more acceptable level as well as being mathematically far more correct than what Sir Isaac Newton did achieve with his guessing about issues he couldn't explain. To be successful in my quest to find an explanation for gravity, I had to redirect all my concepts I previously had and also alter all the otherwise normally accepted thinking on physics. I had to find the phenomena and I had to dissect the function of each phenomenon as well as mathematically valuate the phenomena. In this process I realised that to come to realise what gravity is, I had to realise that gravity is not what Newton saw forms gravity. Newton invalid arguments are easy to see when even looking at the way the man started changing his proof. When the first thought panned out miserably unsuccessful, he then changed his initial

formula that was $F = \dfrac{r^2}{M_1 M_2}$ to $F \; \alpha \; \dfrac{M_1 M_2}{r_2}$. To deceive all thoughts, the man placed

$F = \dfrac{r^2}{M_1 M_2}$ equal to $F \; \alpha \; \dfrac{M_1 M_2}{r_2}$, which is mathematical fraud! The correct mathematical

procedure to follow would be to change $F = \dfrac{r^2}{M_1 M_2}$ in context to $\left\{ \dfrac{F}{1} = \dfrac{m_1 m_2}{r^2} \right\}$ and by

changing the formula by only changing one symbol α he contemplated that the entire outcome of the formula changed without having anything change. ...and then the man is hailed as a mathematical mastermind…no the greatest mathematical mastermind that ever lived! He was convinced by changing = to α it wasn't changing anything. Newton saw it fit to replace ▮ with α and to his mind (if he was genuine unaware of the misconception his actions brought about) the formula was reborn in value while staying the very same. There is an applying rule or law in mathematics that says when

one change a formula from $F = \dfrac{r^2}{M_1 M_2}$ to $\left\{ \dfrac{1}{F} = \dfrac{m_1 m_2}{r^2} \right\}$ then F being F ÷ 1 must also

remove a position to become 1 ÷ F making F the fraction value.

 All those who know even the least about mathematics and of which Newton and his followers know very well, that if any part on the one side changes dynamics from being on top of the dividing line to going beneath the dividing line, then the very same must apply on the other side. One can't just say

that to change a formula, from being $F = \dfrac{r^2}{M_1 M_2} = \left\{ F \ \alpha \ \dfrac{m_1 m_2}{r^2} \right\}$ and such action would

not translate in ultimately changing the outcome of the formula because the truth about mathematics

is that $\left\{ F = \dfrac{r^2}{m_1 m_2} \right\} \neq \left\{ F \ \alpha \ \dfrac{m_1 m_2}{r^2} \right\}$ but when it is required, then the procedure of changing

one, then every aspect in the formula change to $\left\{ F = \dfrac{r^2}{m_1 m_2} \right\} = \left\{ \dfrac{1}{F} = \dfrac{m_1 m_2}{r^2} \right\}$. Newton had

this idea that because he was "Newton the Great (Cheat)" normal rules did not apply and with him being Newton even mathematic laws was below his status. Ignoring laws is the behaviour trademark of criminals including all persons from Kings down to beggars. He was convinced that he could

replace the symbol ▮ with α as in $F = \dfrac{r^2}{M_1 M_2} = \left\{ F \ \alpha \ \dfrac{m_1 m_2}{r^2} \right\} = \left\{ \dfrac{F}{1} = \dfrac{m_1 m_2}{r^2} \right\}$ and

that will change All laws guiding mathematics principles forever. It never dawned on him or on his

followers that came after him that $\left\{ F = \dfrac{r^2}{m_1 m_2} \right\} \neq \left\{ F \ \alpha \ \dfrac{m_1 m_2}{r^2} \right\}$ but the correct application

instead is $\left\{ F = \dfrac{r^2}{m_1 m_2} \right\} = \left\{ \dfrac{1}{F} = \dfrac{m_1 m_2}{r^2} \right\}$. But then he went much further and cheated the

cheated by introducing $F = \dfrac{r^2}{M_1 M_2} = F = G \dfrac{M_1 M_2}{r^2}$.

There was never one Newtonian that even hinted that any Newtonian could explain how the initial

thought of $F = \dfrac{r^2}{M_1 M_2}$ then mathematically changed to $\left\{ F \ \alpha \ \dfrac{m_1 m_2}{r^2} \right\}$ which was intended

to become $\left\{ \dfrac{F}{1} = \dfrac{m_1 m_2}{r^2} \right\}$ and then that this changing was normal, an accepted mathematical

procedure that was in place while protecting all mathematical principles still applying, changing this

lot to $F = G \dfrac{M_1 M_2}{r^2}$ Furthermore, how could academics in mathematical physics teach

children or students in physics this as the truth! How could any mathematician explain a process of following logic, maintain that $F = \dfrac{r^2}{M_1 M_2} = F = G\dfrac{M_1 M_2}{r^2}$ ...explaining it is preposterous!

Be my guest while I put Newton's suggested equilibrium to the test $\left\{F = \dfrac{r^2}{m_1 m_2}\right\} \neq \left\{F\,\alpha\,\dfrac{m_1 m_2}{r^2}\right\}$ Let's first put values to the equation $\left\{F = \dfrac{r^2}{m_1 m_2}\right\}$ and then

to $\left\{F\,\alpha\,\dfrac{m_1 m_2}{r^2}\right\}$ and find out how equal the two can be when real mathematics apply.

$$\left\{F = \dfrac{r^2}{m_1 m_2}\right\} = \left\{F = \dfrac{5^2}{50\times 50}\right\} = \left\{F = \dfrac{25}{2500}\right\} = 0.01$$

$$\left\{F\,\alpha\,\dfrac{m_1 m_2}{r^2}\right\} = \left\{F\,\alpha\,\dfrac{50\times 50}{5^2}\right\} = \left\{F\,\alpha\,\dfrac{2500}{25}\right\} = 100$$

Tell them to show you that 0.01= 100 and how it is possible to achieve this remarkable ploy. One hundredth is equal to one hundred...is that not as magical as only a force such as gravity can be? It is most stunning and what is even more stunning is the number of fools in physics and other smart arses who are unable to see this magic for what it is and that is cheating with mathematics!

Now you try and tell me those frauds aren't cheating when they echo Newton! ...and don't let them come out with this nonsense about figurative speech because there is no figurative in the distance there is between the Earth and the Sun or the way gravity works by measure of a suggested

mathematical formula such as the one our great Newton first used as $\left\{F = \dfrac{r^2}{m_1 m_2}\right\}$ and then later

on when nothing was working mathematically then he went about replacing his initial mistake

with $\left\{F\,\alpha\,\dfrac{m_1 m_2}{r^2}\right\}$. This is mathematical incoherency and is nothing less than committing

mathematical fraud on a grand scale. I challenge you as a student to use this procedure in a mathematical test and see how they blow your paper with noughts and crosses until it is covered in red markings! Where Newton did it, it was alright but when you do it, it is wrong. There just can't be mathematical laws applying to Newton and another set applying to you!

Let any academic mathematically show how one would go about and use Newton's visionary formula $F = G\dfrac{M_1 M_2}{r^2}$ to calculate the force of gravity by replacing the symbols with the actual values in

mass that the symbols should have. Put in the Earth's mass in place where it belongs and put in your personal body mass in the place at M_2 where it should be and then divide that with the distance between your soles and the Earth measured in micro millimetres by the square thereof! Oh yes for a measure of further corruption multiply the gravitational constant with the answer and see how that will lesson the ridiculous nature of the entire exercise. If it can't be done, then that must serve as undoubting proof of Newton committing fraud when he introduced the formula $F = G\dfrac{M_1 M_2}{r^2}$ being able to calculate the force applying as gravity.

It is said that the entire physics rests on this formula $F = G\dfrac{M_1M_2}{r^2}$ but when claiming this please do show how that statement is correct. Take any formula used in daily physics and show where they use the mass of the Earth as a factor in calculating anything. Never, not once, does any formula used by physics hint that the Earth's mass has any influence on any part of physics when any one calculates factors to determine whatever they wish to determine. If the Earth's mass is never used in any calculation, then the Earth's mass has no part presented as a factor and then the Earth has no mass that influences any aspect of physics. That means the Earth's mass doesn't produce gravity because if it did, the calculating formulae used in physics must use the Earth's mass as a factor in all calculations! Newton cheated to bring in the Earth as a factor that has mass that produces gravity and never does the mass of the Earth contribute to any part in any of the many calculations that form part of physics. The Earth has no mass because the Earth's mass never plays a part in any formula. It is as simple as that! The formula Newton first devised has not even a ring of truth to it. If it is true then show how the formula reading $F = \dfrac{r^2}{M_1M_2}$ is used to indicate that this brings about gravity used in any calculations in ordinary everyday applied physics. Even with Newton's further cheating of mathematical principles, show how the use of $F = \dfrac{r^2}{M_1M_2}$ to become $F \alpha \dfrac{M_1M_2}{r_2}$ is legal or where it is used in other forms of formula. Even if any one can show a place where this formula $F \alpha \dfrac{M_1M_2}{r_2}$ is used in a valid legal sense, then please entertain the world with Newtonian insight. By showing that you will be able to prove that I am incorrect of my assessment of Newton's committing of blatant fraud in changing the formula from $F \alpha \dfrac{M_1M_2}{r_2}$ to be then able to function as $F = G\dfrac{M_1M_2}{r^2}$ Even as $F = G\dfrac{M_1M_2}{r^2}$ being in this form, it still doesn't apply.

This clearly shows how what Newton saw as gravity, can't withstand even the slightest test of proof and I showed that it is not possible to use Newton's formula as Newton suggested it applies to mathematically calculate gravity. I come back to this issue later on. I have tested Newton's thinking and the book I offer to you for investigation serves as the testimony to all the testing I did on Newton. This any body who can see, will see when reading this book, I tested Newton from all the angles to see if he possibly could be correct but found his thinking wanting every time. The truth about Sir Isaac Newton's concepts I came to conclude, was that the reality is that it is not in any way overstated to declare that Newton conspired to defraud science and moreover that he committed blatant mathematical corruption in trying to prove the concept he had about what he thought forms gravity. There is no backing for Newton's ideas and even the ideas which are in use are not in the form that Newton said it applies, where physics in daily use serves as the best discredit to Newton bringing no proof about any of the claims that he (Newton) made on matters concerning science in cosmic gravity.

Their attitude of having disregard for all other opinions, is so classic Newtonian. The manner of regard that the Academic Physicist holds to life and the outlook on life that the followers of Newton physics have is quite the opposite of what I think of them. They regard their position as that they are those forming the absolute cream of human intelligence, while I call them plainly Newtonians and to me they are sheepish because they resemble the image that seems to me to be the same as sheep running after their leader, always following without having the ability to think for one second about any thought of why and how they follow their leader or where the leader is going while such examining of the situation is coming from their personal intellect. By having all the power that they do, they have no need to face me or face any evidence I bring.

I want to know how long one length of a measured value is that comprises of and is holding "nothing". Newtonian Master-minds has about 150×10^9 kilometres that fill outer space with nothing all the way between the Sun and the Earth. By knowing what the distance is that one unit of nothing takes up, I then will know how many nothing I could place in a straight line in one micrometer. That would enable me to complete a single straight line holding nothing from start to end in a distance of one millimetre. From that I could calculate what number of nothing would fill the distance in a meter in order to project this number to the distance of one kilometre so that I then could know how much nothing makes a straight line in one Astronomical Unit formed by a single line holding nothing from end to end. If that is the thinking ability that the brainpower of the most intellectual group of humans on Earth shows, then God help us, because then it is a small wonder that about ninety percent of this lot are self proclaimed atheists. The understanding level about reality that these Intellectual Super-humans show, (which are the group forming the ultimate there are in mind capacity and thinking ability only found in the group called Newtonian Astrophysicists) is clearly equal to the understanding my dog has about the reality of practical mathematics and what they show in understanding their surroundings are equal to my dog's perception as to what forms acceptable levels of any sane argument. Fortunately, I am familiar with my dog's level of understanding and his grasp of perception. My dog's equal lack in understanding the basic human ability to argue, places my dog at the forefront of this group of God loathing atheists just because my dog with his mental ability is also an atheist, just as ninety percent of all Newtonians are. The atheistic Newtonians and my dog all have no comprehension to see the presence of God in Creation and neither have the Newtonians and my dog the mental insight to realise that with the aid of the Bible I prove physics correct! That I have done in the book I named "**Seven Days Of Creation**", which is part 7 of **Matters Time In Space: The Theses** It seems they and my dog all share one brilliant mind filled with the nothing they wish to distribute throughout their entire cosmos. No matter how hard any one tries to convince my dog about God, my dog remains an atheist, not because my dog is brilliant, but on the contrary my dog is too stupid to realise issues concerning God. Any one being equally as bright and clever as my dog, has the ability to be an atheist! Are the atheistic Newtonians not a special clan...being atheistic and able in filling the cosmos to the brim with nothing!

The idea of having any idea that the cosmos is filled with nothing shows that the thinking Newtonian's head is filled with nothing because if there is distance in between objects, then something has to fill that distance while nothing removes all distances! "Nothing" is just not capable of filling distances or having measured values except in the heads of Newtonians. One of my books I named **A Cosmic Birth Dismissing Nothing** was rejected because I disputed the validity in the concept that it is nothing that is filling outer space as Newtonians insist on believing. The reasons that was given why I got the publishing rejection by the publisher of **A Cosmic Birth Dismissing Nothing** was said to be on the grounds that Mainstream Physics would never accept my arguments on the matter of nothing not being able to fill a Universe and that I was of the opinion that science has the task to find what forms that that fills outer space with any valid measured value. But those judging the correctness of the content of my book had the opinion that it still remains nothing that fills outer space and since I am incoherent about their view of nothing being unable to fill the cosmos, I had to be mentally retarded. Because I am having the opinion that nothing can't fill the Universe, my opinion is in conflict with the opinion of mainstream physics and that makes my arguments and therefore my book incoherent. I have it in black and white in a letter that one University in South Africa wrote explaining to me as to why they decided to reject the viability of my book. This was what formed the grounds for their view as to why my work is unacceptable and therefore due to that, my book had no commercial or academic basis. What can be more ridiculous than giving nothing a measured value other than zero and if it is zero filling the Universe how much zero is used to form the distance there is between the Sun and any or all the planets! I was subsequently accused of being incoherent in the line that my arguments took, but that is coming from a person or a group of persons who have so much insight into what is possible that they are putting nothing in between the Sun and the Earth while giving that nothing a valued measured distance of 149.6×10^6 km.

Newtonians are most adamant that the Universe outside what is regarded to be material is filled with nothing and nothing forms outer space. Let's quickly ponder on this for a second or two and find out how much of this concept is palatable. They are then saying there is a long line of nothing standing the one after the other where one nothing is following the nothing in front while the nothing in front is leading the nothing behind. The nothing are lines forming rings where every ring ends at a point that the nothing that forms the line of nothing connects, linking in a chain of nothing from the Sun all the way to Pluto and even far beyond, in fact as far as the mind can take nothing and then nothing links in a line even further.

There is a line of thought that the entire Universe not holding material is equal to the value of nothing. It is thought that what links all material (stars and such) is a chain that has a measured value totalling nothing and this will include the gravitational constant which they then say has the value of 6.67×10^{-11}. How it is possible to have a value of 6.67×10^{-11} while also being a value of nothing? It is so Newtonian that I am prepared to admit that it is again above my understanding of Newtonian physics. Outer space is filled to the brim with nothing and that means there has to be immeasurable points where each one of any of the point having nothing is equal to having also the value of 6.67×10^{-11}. That means G is nothing. We have outer space filled with nothing while the nothing Newtonians fill outer space with, hold the value of G or then has the value of 6.67×10^{-11}. Well, the question is where this insanity originates. Of course we find the most intellectual minds in mathematics could not part Newton's trash from the obvious truth and from where does this first thought originate that the universe comprises of nothing...from Newton of course. It comes from the time when Newton said that the spin of an object cancels the space in which the object spins or in mathematical terms Newton said $\dfrac{dJ}{dt} = 0$. That then says where there is space, the space contains nothing. This statement boils down to what Newton argued that when anything spins, then the object returns to the same point where the object started and that cancels all the work done in the space but when dissecting the truth, it says the spinning removes the object from the space it has $\dfrac{dJ}{dt} = 0$. I am fighting the mountain called the inconsistency and incoherency in the world of science in physics. Since no one ever brought a stitch of evidence that supports Newton and by supporting Newton blindly, that brought answers about questionable realities and logic went by the way side since the proof it requires goes unanswered and when employing Newton's ideas without question, it is leaving a void of unexplained facts. This motto of accepting the ridiculous blindly runs like a thread through the entirety of cosmology. Anyone with a doctoral degree in physics can make any blind statement and then as proof back such a statement up only by supplying an even bigger hoax in the form as some grand mathematical formula that is supposed to prove any ridiculous theory. I made a study of these facts while trying to introduce my concepts and this meant that I have been out of earning for eight years while researching the work I now present in many books.

Every book I submitted found rejection because it was in conflict with mainstream physics. After every time that I submitted a book with a different title, proving yet another concept that was never yet explained, I did so in the hope to find some form of recognition when presenting a much more plausible concept than the ridiculous rubbish that science was promoting at present. Every time I submitted a manuscript or sent a book to some University, I lived in the hope of it ending the poverty that this for ever trying to convince all with more research, placed on my family and me. However, every time some inconceivably stupid reason was given for yet another rejection that extended the suffering and the poverty went on uninterrupted for years. I give the answers. I take away from science the stupidity and the incoherent nonsense they so blindly cling to and replace that with sensibility in proven facts but not one academic would look past my criticism of Newton's corrupt thinking!

I wish to bring home the point that it is not because my work is abysmal that my work is rejected and that is the horror part of the ghost story. Every time I presented the academics with more work bringing even more explaining and more proof to my case in more books, I did so while living in the hope that they will bring the recognition that I am due and my books will be published, but then finally the time came when I concluded that this will never happen. They (the paternity of physics world wide) will never recognise my work notwithstanding the accuracy I present. By doing that they will create a monster they can't control while the monster Newton controls them. All I need is one academic, even retired, who admits my work has merit and my work shows more correctness than that which the Newtonian ideas at present deliver.

However, no sooner does that academic agree to my concepts than that one who recognises the merit about my work becomes the first academic that denounces his or her personal life achievements. I am not disputing the Big Bang concept, because finally and at last Newtonian science has achieved some correctness after many centuries of mismanaging physics, but I surely can explain a lot more about the cosmos than pushing everything on such simplicity as fathering mass as the only inspirational concept there is in the Universe. Slowly I realised I had to turn my back on academics because they will never befriend my views. Now finally I came to realise that if I wish to succeed in any form, it would be by ignoring the academics just as much as the academics rejected my opinion. In the same way I now reject their opinion. I wrote to all the established Universities in South Africa wherein I promised the heads of those departments a fight and with that last letter in October 2005 I severed all efforts to establish any form of acceptance from academics.

To me now, in the present, they are as dead as I was to them when I tried to convince them about my work. I now charge the public to put their wisdom in place and use public opinion to put those arrogant Academics on trial. I wish to publish a book in which I can show in the presence of the general public just how stupid they truly are. I now wish supply the students with questions that students may use and that will uncover their deception and I hope the students will make them look like the fools they truly are. Here are a few ideas you might ponder on in order to judge their mentality.

While we all know and admit that the cosmos is never-ending and therefore I can say that everyone accept that fact about how eternal the Universe is, Newtonian scientists have put edges and limits in place in the Universe from where Newtonians can discover undiscovered things that previously was never yet discovered. Visit the internet and see how many discoveries cosmologists locate right on "the edge of the Universe". The limits they place on the Universe are not the limits they claim there are, but it is placed in order to hide their personal insignificance. That nullifies the claims that Newtonians as scientists understand the Universe they investigate and the reason for that is that they have the entire basics of Newton's formula holding and explaining the Universe as being

$$F = G \frac{M_1 M_2}{r^2}$$ completely wrong.

If everything is in contraction, then by now some places should already contracted large areas of space leaving gaps at other places where cosmic holes should by now be in place. If contraction brings about the gain of space in some parts of the cosmos as Newtonians say, then there has to be a part of the Universe that is losing space and by losing space a shortage of space must bring holes in space to appear where space is reducing! The Universe has to be in a balance to progress by some order while ultimately stay in balance and maintain absolute conformity. I have put this in a book again named as a letter which holds one thousand two hundred A4 pages and was told by iUniverse that the book is too large to be printed in its entire form in a Print-On –Demand format. I then decided to keep it back and move on to try and get other less informing books in published format. In that book I use all the information from other books combined as **Matter's Time In Space: The Thesis** to show how this balance is conducted and how does the Universe grow without ever increasing anything ever. Space is not increasing because the Universe holds whatever there can be and therefore space or growth can never increase what ultimately is everything, which is called the Universe. The balance is in material / space and time and the one is a part of the other, where displacement or movement is the combining factor. It is a matter of finding a balance between that which can never reduce more, since it represents infinity, which is the smallest anything can ever become and in balance to that is eternal – everything, which will forever increase and that is because it represents eternity, or eternal movement which seems to us as eternal growth, but it is not eternal expanding. I show why space seems to grow and yet while it could never become more than it is, as it represents everything there already is, but explaining it is not that simple as to just mention this method of space increasing or Hubble growth since by not motivating and substantiating everything I may claim, doing so will degenerate me down to the levels held by Newtonians at present. In short, if the Universe is everything there already is, how is it possible that everything that already is there then could become more? It just can't become more. Therefore what we see as growth of space is relevancies shifting the balance that is holding the Universe disciplined, but to understand this process in detail, one has to see Newton for what Newton is!

I realise accusing Newton of fraud is controversy where the Da Vinci code is not truly controversy. This work that I present is the biggest controversy ever written, where I take on the might of the Academics in Physics in what they present as science and I take on the might of everyone who is an academic in physics and those who are presently enjoying the total power the department of physics has to offer. In the book I hope to have privately published, I for the first time ever, manage to prove what gravity is and I prove why gravity is there to begin with and moreover that it has no similarity or any common concept with anything Newtonians envisaged or with anything Newton ever said. But when that is said, it then is also not being what Newton said. Any confrontation with official physics policy leads directly to a situation where academics shoot me down in flames and with the negative advice given to publishers, my work is rejected.

I have sent offers to view my manuscript to eighty publishers at one point where I prove, by using physics, that the beginning of Creation was exactly like the Bible says and according to physics it happened that way to the letter and the word. If you change one word about the events in Genises1, physics and the Bible is not correct. I named the book **Seven days of Creation**. I had the same conditions applying that I also put in place with you at a publisher where I demand total confidentiality given specifically from the publisher before I send the manuscript but that is to protect my vulnerable situation and I explained that in the previews I sent. This letter, for instance I consider or name or think of as a pre-view. I truly explain how the Bible is correct about the birth of the cosmos and that I do with applying physics and with physics the way I prove gravity works, I can and do show how events happened as the Bible describes the Beginning. I got not one request back from any publisher to read my manuscript. It was then, after that episode that I started to turn against the might of academics in physics. But even then, after that episode with **Seven days of Creation** being totally ignored, I was still not even remotely as aggressive towards academics as I now am, but that is where my negative attitude and confrontational demeanour started.

With the unlimited power given to them, academics in physics can avoid confronting me because they can afford to merely dismiss me as if what I present was never presented and they don't have to prove Newton is correct because that was never done and disproving anything that was never proven is quite a task, because the acceptance thereof became culture and religiosity. The correctness of Newton is accepted although Newton was up to now never proven correct. Newton suggested that mass is responsible for establishing gravity in the face of contradicting the already accepted Galileo principle and the fact that mass brings about gravity or the later imported idea of the existing of a "graviton" is merely wild oats sown by suggestion and the suggestions are accepted for the past three hundred and fifty years. With me going against the grain of Newtonian society, I am going against the grain of everything ever accepted in science and science would rather dispute God Almighty than they would dispute the deity called Newton.

By proving for instance that mass has nothing to do with gravity but gravity has everything to do with mass (just to name one aspect), I show that mass has no function and has no purpose in the Universe except to make Newtonians feel well informed and allow them to think they have some control over that which no one can ever control. If mass was valid, then the strongest and most massive there can be would not be a Black Hole because a Black Hole is that star that essentially has no mass at all and with having no space the Black Hole must qualify as being very small, that is if size brought about mass. With no space, the Black Hole must have no gravity when applying Newton's standards on size, and yet the Black Hole has absolute gravity. Yet the gravity a Black Hole employs unites time and ends the Universe at one point. All this becomes apparent when reading my work and it is so simple to understand since I have a simple mind and therefore have to use simple methods to explain how I see the cosmos unfold mathematically. In the books however, I had to touch on the most important issues, keeping the concept intact and doing that also became the reason why I had to split the one book **An Open Letter on Gravity** into the two books they now are, because there is a full compliment of facts I have to present to substantiate my reasons that validate my arguments. I had to maintain the informing basis of the books as to keep the integrity of my effort in representing the truth.

Since 1977, the year that I finally ended my studies, I was convinced that there was something amiss in the approach science took on the matter of gravity. Newton's assumption that $\frac{dJ}{dt} = 0$ is possible was completely incorrect and the confusion that such a presumption caused, rubbished science totally. The fact that Newton believed that dividing anything into anything ($\frac{dJ}{dt} = 0$) and ending such a division with an answer of nothing that presumption Newton made did not sit well with me. I studied the discipline Johannes Kepler introduced in much more detail than what Newton tried to have the world believe and the Universe I discovered using the discipline Johannes Kepler introduced which is space $\mathbf{a}^3$ is equal to $= \mathbf{T}^2$ movement $\mathbf{k}$ in relation off $\mathbf{k}^0$ singularity or $\mathbf{k}^0 = \mathbf{a}^3 \div (\mathbf{T}^2\mathbf{k})$ where $\frac{dJ}{dt} = 1^0 = k^0$ which then puts Newton's thoughts in line with Kepler's thoughts being $\frac{a^3}{T^2k} = k^0$ and then correcting Newton to how it should read corrected the entire view of the Universe! I based my theory on the discipline Johannes Kepler introduced. I rejected $F \;=\; G\,\dfrac{M_1M_2}{r^2}$ as complete rubbish!

Because I am of the opinion that there is a large principle incentive not to get my work published, I don't for one minute suggest it is organised or pre-arranged but every time I try to get some book published I walk into a wall of resistance coming from some academic corner. I don't say there is an organised conspiracy formed to plot against me (fortunately I haven't gone that mad yet) conspired by Mainstream physics and spear heading an onslaught directly at me, but it is a mentality amongst academics to reject my opinion (or any other opinion that disregards Newton) when I attack what Newton claims, whether my attack is in introducing another view not associated with his opinion or when using a much more direct attack such as the one you are reading. Newton stands protected by a system that is in place and the academic world holds Newton in place notwithstanding the total absence of proof and with no scientific accuracy used as a basis. But since every paper that has been written was about echoing the view Newton stood for and this huge industry would go tumbling into the gutter where they belong, it is therefore that academics reject my work without reading or testing my views.

Because everyone this far has based all work they did on haling the perception as to what gravity is in accordance with what Newton's ideas perceived gravity to be and how gravity comes about through mass establishing gravity, the Masters in physics blocked the publishing of my findings on every conceivable level. In order to overcome this barrier, they used to block me intentionally or otherwise, I went about writing a book with having in mind to publish my book privately by using a publishing principle called "Print–On– Demand". This too I found much tiresome and unbelievably complicated to achieve. My work is not the average Mills and Boone easy to print layout and it requires more than the average input to achieve the setting due to the pictures I use to portray some new concepts which I have to introduce in order to get my work understood.

I know the sketches make the life of the publishing jockey miserable and that they opt to get out of the grinding stone, but there is no other way any person would understand my books than to use pictures and sketches, as you will see when reading the work yourself. I found this to be the case when I first submitted work to a firm called iUniverse to have the work printed on a "Print–On– Demand" principle. It was the only away forward which I saw and which I still see. One thing about those jockeys in the printing press at iUniverse is that they hate back-breaking work.

In order to try and promote my work, I sent a book away for printing. The printed to be book I named *__"an Open Letter to Selected Academics"__* . This was an elaborated version of the first book which I gave an I.S.B.N. number and of which I sent eighty copies to eighty Universities world wide. I sent the books in the hope of finding only one academic who would see what I proposed and then would be willing to promote my work as an academic. Before this venture I have tried selling my self published work on the Internet, but a book such as mine is buried so deep under a pile of other work, that my book will not see daylight for another million years to come.

Then I wrote this book, which was aimed to be published on "Print–On– Demand, but my work is extensive and seemingly complicated. In that time it also dawned on me that the book I submitted and which I named *__an Open Letter to Selected Academics"__* was directed at academics that were unwilling to read my work even when they received such work free of charge. If they refused to read my work, which they received for free, what on Earth will motivate them to buy my work they refuse to read? Realising this mistake and in the light of all the publishing problems I encountered, I withdrew my work from the publishing process and again came into contact with academics in the hope one will relieve him or herself of the state of being fixated with Newton and just read my work. Condemn my work if I am wrong, but then show me where and why I am wrong when I present any of my arguments.

Then the idea dawned on me to get confrontational and aim directly at the public whereby I would inform the public and moreover students about the unacceptable affairs going on in physics. The work included and even starts with the final confronting letter I wrote to the nine Universities in South Africa. The book was aimed to reach the public and to inform the public about my principles and not that much about condemning Newton.

This book I named *__an open letter Announcing Gravity's Recipe.__* I thought that once I had the book to offer, and I got you to show how unreliable Newton is, it would advance my book's selling potential. I paid a firm going by the name of Cork-Hill publishers to do the publishing and was met with a lot of enthusiasm about the venture. They seemed very sincere and had a lot of enthusiasm at the time about the project. I paid then $1000 to get the work in the press and sat back waiting for the book to emerge from a manuscript to a published article.

During the time I was expecting the book *__an open letter Announcing Gravity's Recipe__* to be published I set about doing the framework for a book with a publisher in mind. While doing the framework which was not intended to be a book by any standard or was never thought to become a publishing work as such, I had the idea which I thought would become a solution to the problems I encountered at iUniverse. If there were some demand coming from the selling department for the book, it would put some pressure on the printing department to get the book done. By completing a framework book I was hoping to interest the program management of a publisher in airing my views on gravity while introducing the book *__an open letter Announcing Gravity's Recipe.__*

This framework I set up was however a much abbreviated affair and my intention was that it was to be used as some framework that only had the necessary information I thought would interest a publisher. I intended to bring to the attention of a publisher very little of the information I had published in *__an open letter Announcing Gravity's Recipe.__* I thought that a publisher aired the content of the book *__an open letter Announcing Gravity's Recipe__*; it may evoke a response from the public and if there was some degree of public demand to purchase the book it would motivate the publishers in completing their task. The framework was written with a principle goal as to contact a publisher and informing a publisher about the information contained in *__an open letter Announcing Gravity's Recipe__* and in view of that I hoped that a publisher may have an interest in the content for the purpose of airing it in a program. The framework was meant to keep me busy until such time as the book *__an open letter Announcing Gravity's Recipe__* was on the shelves of book outlets. By March 11 2006 which was about four months after my payment was made, I realised I have been conned and that Cork-Hill – Press had no intention of delivering on the promises they made and if I wanted a book published I had to do another one. I also realised while working on the framework meant for a publisher and which I referred to as *Newton's fraud* just to get some title going at that time I realised that even the book ***an open letter Announcing Gravity's Recipe*** was written with finding some academic response and the true commercial incentive is still left on the back of my mind in that publication. I was out to impress academics and not to inform the public as I intended to do.]

This I realise when a friend of mine asked me on the date I mentioned to explain to him in layman's terms what gravity is and with both of us being layman I could not use mathematics as the main tool to bring across my arguments. I had to use plausible language spoken by all and with that I realised I was still tampering with academics' thinking even in the book I was supposed to stick them with an attacking sword. From that I set up the a publisher script and developed the three part (one having two volumes) I subsequently named **An Open Letter on Gravity** and in doing that then the framework I compiled for a publisher got a body and received much more explaining than the framework initially intended to have had. The framework was never intended to become a book but it now has in the form of the title **An Open Letter on Gravity**.

This book rose from the framework I initially intended to serve as a guideline used by a publisher and using another name for the title in the program intended to be aired by a publisher. On this matter concerning the Title I named it as ***Sir Isaac Newton's Fraud***, or another name I thought to use was ***Sir Isaac Newton's Fraud: A Conspiracy to Defraud Science*** but it was never at first intended to become a book while it now has become a book. On this part and how this happened as well as why this happened I elaborate more at the end of this introduction.

The book that developed from this book I combined with Cart Blanche in mind is now a book but it is not part of the book holding three books with two parts and the part 1 has two volumes. The book I now offer a publisher is a development after I completed the three books entitled **An Open Letter on Gravity Part 1 (volume 1 and 2)** and **An Open Letter on Gravity Part 2.**

I named the resulting book **An Open Letter on Gravity Part 1 (volume 1 and 2)** as well as **An Open Letter on Gravity Part 2,** and those books are still intended to become published by the principle "Print-on-Demand". It would be much to risky for a publishing house to publish the work without having some academic blessing or some degree of academic support and that it will not get in the climate we live in at present. But even in having the work published on the principle of "Print-on-Demand" it has many drawbacks and incentives not to be printed.

The layout of the work is most difficult to do with the many sketches added in the work. I have submitted a manuscript but the firm was not satisfied with the form it had at the time that I sent the first manuscript away. I had to redo the book because the volume of space it carried was too large for my computer to handle in one volume and with the limited capacity I had to split the book in many parts. This the publisher found to be unacceptable and then I had to rewrite the entire book to fit three parts or I had another choice which was to reduce the book to fit into one manuscript. This was no solution because by dong that it meant that I had to remove some information and as you will see if you wish to read those books, all the information is vital in supporting the arguments. The sketches are vital in explaining the new concepts and without the sketches the work will be poorly understood. To put my case I have to present certain facts to the reader that makes my case viable and understandable and by reducing to work I present, I weaken my arguments entirely.

I bring a solution to science, but as my solution does not match their vision or the work they submit as being what is acceptable and is correct, their rejection comes even before reading my work. If only one would agree that my work has a basis, it will change science completely and that in effect renders every book and every thesis ever written on astronomy to be total rubbish as far as scientific substantiated fact goes. This is a long story but at the end of the introduction you will realise why I do all this explaining about a book I do not intend to be used by you. The book I am referring to is named **An Open Letter on Gravity Part 1 (volume 1 and 2)** and **An Open Letter on Gravity Part 2** while the book I submit for your use is named either ***Newton's Fraud*** or ***Sir Isaac Newton: A conspiracy To Defraud Science.*** The first mentioned books all carry my new concepts while the last title only uses principles that Mainstream Science uses and by only showing the incorrectness of Mainstream

Science, I do not have to involve the opinions of academics because in the last book I challenge the very principles those academics support.

I was able to prove the Universe is a sphere, and this simply means that is why the Universe can never end. In the past we presumed the Universe to be a sphere but it was never proven mathematically. Other scientists with greater minds never even considered this as an issue but this issue is a pillar on which gravity rests. Why is proving a sphere important...because only a sphere has the ability to establish gravity, because it is the foundation of the curvature of space-time!

I was able to explain all the phenomena science found in the cosmos but science were never able to understand or to incorporate these phenomena in Newtonian science, so Newtonian science dismissed / ignored the phenomena. Those phenomena science denounced, as facts are what form gravity and that much I prove mathematically. While those four phenomena form the pillars on which the cosmos rests, those four phenomena never supported or proved to use mass and because of the absence of mass, Newtonian science dismissed them as being either coincidental or disputed any validity thereof. Yet, some was used to detect the location of undetected planets in the past.

I prove their validity as a matter of fact in the book I offer and that is how simple it is if you stick to the simplicity and not run away getting technical beyond the human mind as most Newtonians wish to do. The detail I use is comprehensive because there is always cross-referencing in every line I wrote in the books. Without the reminders supporting the argument, the argument goes lost and becomes futile. That makes the book seem comprehensive.

These books are the first in science ever to uncover what I claim that I uncover. It takes science head on by battling against the form in which science at present thinks and the fight is on a head-on collision. With this I say that the arguments in these books do not beat about the bush presenting the discrepancies we have in science at the moment. I have to show cracks in Newton's armour and show the world of his failures in order to set the world free from this Newtonian idea that Newton is infallible.

You can imagine what the selling potential would be of a book when any book lands on the market and that book can reveal what I claim my books reveal. Think what selling potential a book may have in revealing what gravity is and what is the potential selling ability of a book that can show where the centre of the Universe is or how many copies would be bought if gravity as a factor is explained in a book going on the market. With your help I might just find the means to put the book in the hands of the general public.

If you would care to contact me, I will provide the manuscripts that contain my personal work and then you might take the opportunity to read all about my work. You then might be able, with the information you obtain from these books I present you with, to judge for yourself how serious my arguments are. The work I present is a drop in comparison with the work I already wrote and what is waiting to be published. Where I did present my work to institutions, they rejected it before reading it because it clashes with Newton head on. With my views contradicting Newton, it is swept under the carpet and they did that so many times before with other work I did. Every time they found new idiotic principles on which to reject my work. Reasons given in the past for rejecting my work was that my English is not up to standard and of a very poor quality, I suppose because I have a name being Peet Schutte and more Afrikaans than that is hard to find, Newton was never yet proven wrong, and any attempt in doing so has no validity, that the arguments I put forward are incoherent and not matching general accepted science, also that Mainstream Physics will never agree with my arguments...and the list goes on and on...On this occasion I saw to it that the books I wish to have published are not written for academic consumption as all my other work are, but is an abbreviated explaining of the most important issues or the golden thread that unites all science principles.

Presenting my work to the public will raise public awareness of how the Academics mismanage science. I hope this awareness will result in the public starting to put pressure on those, that up to now could get away with fraud because they found no one challenging their thinking and there was no one yet that put true pressure on them. With the pressure of the press and public having insight to their affairs I recon then they will have to come out into the open and explain what they can't explain. I know the work is intensive. Everything Einstein alleged, I prove and I prove so much more.

In respect to the normal channels one can follow or what there is to follow in Academic circles, I could write by launching papers on the issue I address. This is the method, but which it is normally done, and which the route even Einstein followed at first is but when one confronts Newton, that simple writing of a few lines on a few papers does not work because I disagree with most of science. To prove my disagreeing, requires lots and lots of papers flowing into one compliment as a book that becomes thirteen books that I wrote this far. How does one explain that mass does not bring about gravity as was believed since Newton, but mass is what frustrates gravity and hinders gravity and uses only two or three pages to do that?

The concepts are too massive to have them covered by normal academic papers published on this or on that topic. To get the Academics to review my work as a meaningful project with the required enthusiasm and honesty and to cover me from their anger in the process I need a book firstly to disprove Newton from where I then can launch my work by which I can prove my case. I can only get the truth out into the open with the published books and as I will explain later on, this is where I hope you will bring support to my case. At present the academics I approach don't read my work or only read some of my work and that I know because they can only quote from the first twenty pages or so of the work I hand in and never offer any direct opinion about what I wrote. I have no backing and they do not even consider me as being alive and breathing and forget that I may have the right to an opinion, let alone to be so opinionated that I have the audacity as to challenge the validity of Newton.

You should ask the world of science how much attention would be raised when proof comes about that Newton is failing in his concepts on gravity. Think how many students disagree with Newton but are forced by physics to accept it or find another faculty. My work gives me copyright on physics and that is no exaggeration. Writing the framework was on the table but it was planned to come into practise only after a book was published. But I had to bring my date when I planned to contact a publisher much forward because of a personal issue that developed in the meanwhile.

I then set about and started developing the framework I mentioned I initially put in place for a publisher to air into a book holding the Title **_Newton's Fraud_** or whatever and this was done. You have to agree that if the book is truthful and you air such a program that shows all students how they are brainwashed and mislead and on the bargain supply them with questions to ask their Tutors as to place their Tutors in a predicament, those Tutors surely deserve to be in, such a book must have a giant selling potential world wide. Some publisher can make himself a fortune and save the world from academic corruption while raking in money. It just has to be a money maker if I can convince you to convince a publisher that this book only serves the truth as far as physics goes. This is the biggest motivation that drove me at the present to contact you and offer you the book entitled **_Sir Isaac Newton's Conspiracy to Defraud Science_** or **_Newton's Fraud_** or whatever the name of the book is you will choose is going to be called. The sole motivation behind this idea of getting **_Sir Newton's Fraud_** published is the urgency to get a publisher funding.

As you will see the book I offer as **_Sir Newton's Fraud_** is not edited and is lined and riddled with all types of errors. It was written with only haste in mind and was developed in one month from forming as framework to being a book that is showing what science truly hides. If I edit the book it will end up as a book holding seven hundred pages of unedited work and if I edit it after that it will still be unedited with the content doubling again. I have written thirteen books and in that way the one developed from the other as the first one grew too long for use while they always remained unedited. Then we (my wife and I) decided that the editing will only be done by my wife after the final pages are written and at such a time I am not allowed to go any where near the book being edited again. But at this point I have no choice as to present a book to a publisher unedited since the urgency to find a publisher is huge. The facts I put forward concerning Newtonian science are real and you can test the facts using any criteria, but be warned that if you use the help of academics you will find the same resistance I encounter, with the only difference being that they can wipe me off the table but you have a world class program that will scare them and keep them truthful but only if you press them to remain truthful and honest. I am prepared to confront any academic on any issue that I address in any of my books not holding back or discarding one word I said. If you wish to spend the time, you can read the book as copies are supplied and see how I come to conclusions that go beyond further arguing or contesting. I am leaving a copy of the book in Johannesburg for you to collect which you may use to control the content and check my facts in regard to the mistakes and flaws found in physics. The book I leave has only the work in regard to Newtonian physics and has no information regarding my theory. I am prepared to supply work about my theory on a condition since I have to protect my work at this stage from academic theft.

I am supplying you for your possible use or interest a copy of ***Newton's Fraud***. Should this matter interest you further I can provide you with my work as to show I do have the answer if the academics wish to listen. With further contact I will provide you with one copy each of ***An Open Letter on Gravity Part 1 Volume 1 and 2*** as well as two books holding one copy each of ***An Open Letter on Gravity Part 2,*** also in two parts. The copies of ***An Open Letter on Gravity*** are supplied for your investigation and (should you wish to read) I also include those for your convenience as to see that I give answers and not only cast doubt. The book ***An Open Letter on Gravity Part 1 Volume 1 and 2*** are the two manuscripts I hope to publish on a "Print – On – Demand" basis through a publishing firm specialising in such a matter. However I do not intend to offer them for publishing through the normal publishing channels since that venture will be risky at this stage to any conventional publishing house. I have compiled a book above and beyond the one I am leaving with my brother another book that I named ***Sir Isaac Newton a conspiracy to defraud Science*** especially for your convenience in order to show the most basic and crude basics that my theory proposes about gravity. However, with the information released in this book, I am only willing to hand that book over on a personal basis to one of the presenters of a publisher and where I will receive the reassurance that the information is safe from academic theft. It is the book ***Newton's Fraud*** that I hope to find a publisher.

The Book I include is the manuscript that was intended for a publisher and only delves into Newton's failures. It contains no new work that I bring into the open. I have another book exactly as the one I

am sending you with one additional chapter in which I introduce gravity taken from the point where I correct in a simple manner Newton's mistake. That book I will supply but that will only be on a personal level since I am very wary of academic theft and the solution is so simple that any body who reads it will feel like kicking their behinds. But once this goes out and becomes stolen by any academic of stature, I don't stand a chance in recovering my work again. If you feel a need and when you contact me via my brother, I will supply you with my other work I named ***Sir Isaac Newton; a Conspiracy to Defraud Science*** in which I divulge the principle of gravity. Also I have printed and which I plan to supply you with copies of ***An Open Letter on Gravity Part 1 Volume 1 and 2*** as well as ***An Open Letter on Gravity Part 2.***

Newton put $F = G \dfrac{M_1 M_2}{r^2}$ as the basis of all physics all the while presuming that gravity comes about when one object that is supposedly holding mass is pulling another object also having mass and this having mass produces a force of attraction between the two objects, which would encourage the two to move towards each other by the value of individual mass. The entire basis of all physics rests on this formula $F = G \dfrac{M_1 M_2}{r^2}$ just in the very same way as Newton introduced the concept where in it is believed that there has to be mass and with such individual mass, the factor arising from the individually measured value of mass, from such a value a force arises and this produces all gravity.

Such gravity then can't have a constant since the mass function by distinction of differentiation in density as well as size while such presumption just has to exclude g = 9.81 Nm/s^2 because g is a constant at a value of 9.81 Nm/s^2. If physics is anything to go by, then what ever is proven, such proof must stem from and be in support of as well as being supported by this formula $F = G \dfrac{M_1 M_2}{r^2}$ exactly as Newton said it worked. Taking Newton to the letter it is believed that this formula $F = G \dfrac{M_1 M_2}{r^2}$ is what keeps the entire Universe in place. Saying that, then it also must be true that all of Newton's accuracy solely depends on $F = G \dfrac{M_1 M_2}{r^2}$ as a formula that has to be truthful and unquestionably accurate.

The mass is the crucial factor because the mass is taking the position where the mass creates a force according to the value of the mass and according to the individual value of the mass, it creates a proportionate force that is able according to the measure of the mass, to destroy the distance in space of the radius forming the space between objects with mass and this happens according to each value in mass from both ends but also equally from both ends.

On this rides the entire future of all man and all beast and all vegetation and atheist believe in the outcome of this $F = G \dfrac{M_1 M_2}{r^2}$ and this alone, more than they believe any prediction that the Bible may offer.

According to Newtonian religiosity this is the absolute most critical formula ever discovered by man because it is not only man depending on the outcome, but it is the entire future of the Universe that awaits the conclusion of this formula $F = G \dfrac{M_1 M_2}{r^2}$.

With so much at stake and so many Oh-So-Wise believing in the infallibility of this formula, then we also must accept that the formula $F = G \dfrac{M_1 M_2}{r^2}$ has been tested and proven so many times over to a satisfactory conclusion, never delivering any doubt at any point ever, and the accuracy tested by science shows unwavering that there is no other formula on Earth that has endured the testing that Newton's gravitational formula $F = G \dfrac{M_1 M_2}{r^2}$ has under gone. Every atom stands in testimony to the truth and undoubting accuracy that this formula renders. Not one stitch of evidence was ever brought to implicate any contradictions about this formula, even in the smallest sense!

Tell those members of the Brainy Bunch who advocate the fact it is mass that is pulling you down and that it is mass that creates the gravity hat allows a body with mass to fall and ask them politely but also firmly to please explain while they are persisting in using Newton's gravitational formula $F \ = \ G \ \dfrac{M_1 M_2}{r^2}$. How is it possible for a wind that is a hot air forming a draft, to pull a hanging body with mass up into the sky and have a person surge through the sky like a winged bird does...and all this is done while riding on a hot wind. Does the body lose the mass in order to find it possible to surge in the air while riding on a breeze of hot air? What has hot air got to do with the force of gravity and how does hot air prevent mass pulling you down. How can hot air jeopardize the force of gravity when Newton stated categorically that mass pulls the body by the force of gravity? Locked in these questions about lift gained from heat relations changing, we find the answer to gravity and not in the magical mythology of mass pulling by force to create madness! If mass was pulling, then hot air has to be a form of anti mass or if gravity was about pulling bodies down then lifting bodies up must be anti-gravity.

Blowing hot air into a balloon increases the space of the captured air that is inside the balloon. As it increases the ratio in density there is between solids that is supposed to sink with gravity and liquids that has the solids float suspended in mid air, the increase in space gives the liquid (air) more of the advantage in ratio. By filling the bag, it takes up more volumetric space in the confined ness of the balloon as the size of the inside of the balloon bag grows. By inflating the balloon, such filling of the balloon can't be filling the balloon with more nothing and therefore the filling has to be holding something that is worth a ratio change coming about, since that which fills the balloon pushes the entire balloon with bag and cargo higher into the air. The few molecules holding gas elements contained by the air that goes into the bag while the hot air is accumulating inside the bag, can't extend the size that the bag increases that much either, so the air coming in has to be something that takes up and holds space, which is something that nothing are not capable of doing, if nothing is supposedly filling the bag. It is obvious that the air to space ratio increases in favour of the air but the significance is that the balloon lifts and the process neutralises mass attracting by gravity (if mass does attract by gravity).

By blowing hot air into the balloon, the balloon increases in size as a result of the air increasing, thus increasing the confined air and as a result, the confining of more air increases the volumetric space in size, but also this practise counteracts what Newton suggested gravity is suppose to do. Then hot air forms anti gravity because the balloon lifts its cargo of mass. The bag in size increases in volume with captured air which is the confining volumetric space that increases being in relation with the space of the material and is overbearing the unchanged volumetric space that the material holding mass claims. The solids in size and volume remain unchanged when the air bag is inflated by adding more hot air. The existing air to material ratio increases in the favour of the air and transversely affects the overall density of the unit as a whole. Adding hot air to the balloon bag changes the relevancy there is between solid and liquid, where the bag and its content are the solid and the air surrounding the entire balloon (bag and all) forms the liquid component. This brings to mind the operation of the phenomenon we call the Coanda effect.

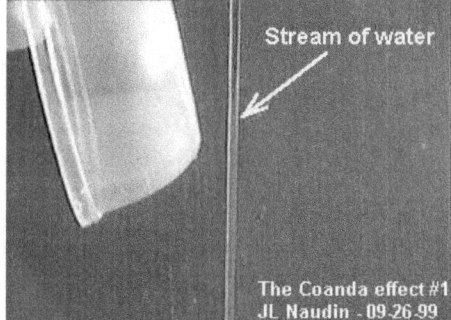
Stream of water

The Coanda effect #1
JL Naudin - 09-26-99

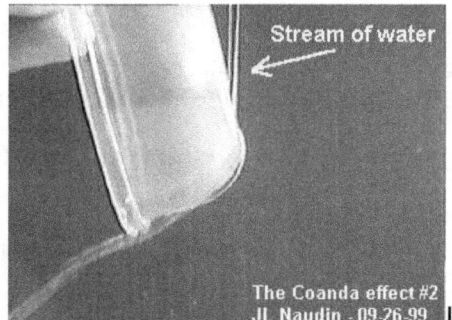
Stream of water

The Coanda effect #2
JL Naudin - 09-26-99

Informed circles such as engineers regard the air as liquid when doing calculations and this is done contradicting the fabulous Brainy Bunch that instead as cosmologists regard air as the continuing of nothing extending from the Earth into outer space as far as understanding of nothing will permit the Newtonian to not understand the complexity of nothing... The increase of hot air into the balloon changes the ratio there is between the solid, which is the bag and the cargo as well as the balloon's material in regard to the captured air size increasing. It is the total density increasing by adding volumetric space of the total compliment of the balloon and the bag as well as the cargo it carries, which represent a density recline in relation to the liquid that the air represents in the balloon. If mass pulls by gravity then by adding hot air to the balloon, such adding of more hot nothing brings about anti mass that must evoke anti gravity since the direction of the balloon changes in the way it travels. This shows that gravity is density dependent and the increasing of captured space in ratio to free space pushes things up while this also shows that there is much more complexity to understand than Newtonian simple minded thinking that mass pulls things down. By the changing of the air ratio in the balloon this shows the density decrease tips the directional flow of space in favour of the liquid by taking the solid up.

This questions Newton's force of gravity that has the mass creating a force that would generate the gravity whereby the pulling of the other object orbiting and in having mass then therefore is also generating mass whereas by mass the other object will also force gravity onto the first object $F = G \dfrac{M_1 M_2}{r^2}$. What goes up must come down and from that perception the proof about this formula $F = G \dfrac{M_1 M_2}{r^2}$ is derived.

We fight our mass because we fight gravity the entire time during one life span we live through. That is the proof that we attach to this formula $F = G \dfrac{M_1 M_2}{r^2}$. When I jump, it is the force of my mass that generates the gravity by which I pull the Earth and by which the Earth pulls me back and the pulling is the result of the mass of the Earth that pulls me down again while at that moment I am pulling the Earth up again, thus the square value coming about in the radius factor and my experience about that is all the proof this formula $F = G \dfrac{M_1 M_2}{r^2}$ requires. That is the entire proof that Newton's argument relies on and finds backing of such a statement. When I fall, my mass kicks into action and by mass I hit the ground at a rate my mass will determine and with that, it bring the absolute proof that Newton $F = G \dfrac{M_1 M_2}{r^2}$ was eternally correct.

Newton is a genius because Newton realised all these wonderful happenings when he put into place $F = G \dfrac{M_1 M_2}{r^2}$. Newton saw that a planet pulls another planet by the gravity that the mass of the planet charges and still the unwavering accuracy about $F = G \dfrac{M_1 M_2}{r^2}$ never came into focus.

Never once did any of the Brainy Bunch realise that with the formula $F = G \dfrac{M_1 M_2}{r^2}$ being undauntedly accurate, then why are there still comets left? With mass pulling mass onto mass and into mass by gravity destroying the radius, what would prevent all the comets from self-destruction as they hit the Sun in an unpreventable collision, well unpreventable according to $F = G \dfrac{M_1 M_2}{r^2}$?

$F = G \dfrac{M_1 M_2}{r^2}$ seems simple and well proven when Newton's arguments $F = G \dfrac{M_1 M_2}{r^2}$ remain on Earth, affecting our situation and our conditions we live in, but let $F = G \dfrac{M_1 M_2}{r^2}$ roam a bit to other planets and one can't miss the contradiction shouting and glaring at every sense of accuracy you may have. However, the truth is not fully applauding Newton even on Earth!

Consider that if mass is what brings about falling, it then implies that objects just cannot fall equal but have to fall differently and according to their mass.

By blowing hot air into a confined space, this hot air increasing can affect the object in mass having a change in gravitational directional movement. Does saying that not ring a bell in the minds of the Brainy Bunch and light some spark amongst the blackness of incompetence forming in the murky fog they have in their minds that prevents their intellectual thought from inspiring any intellect?

When it is true what Newton said about mass being the creator of gravity and gravity is the force of attraction and this lot is there serving one purpose and that is to diminish the radius there is that will have you go falling to the ground as fast as mass will allow, then how is it possible to pump hot air into a canvas, and have cargo float through the atmosphere while hot air pulls the body upwards notwithstanding the mass with the gravity pulling you down? How can hot air prevent mass from functioning and how can hot air form anti gravity that pushes you upwards?

Only if mass has nothing to do with the falling will it be true as Galileo proved that objects must fall equal and in equanimity throughout the entire distance of travel while in the process of falling and that fact goes without argument. Mass brings variation and differentiation between objects and falling depends on conformity while the process of falling is the result of equality because all things fall equal. Such equality applying during the fall is the result of mass not applying! Let the academics put that in a pipe and smoke it!

If Newton is correct, then the Universe must be in a state of contracting $F = G \dfrac{M_1 M_2}{r^2}$ because that is the only conclusion one can come to when going according to what Newton's formula indicates. The objects are drawing closer to each other all the time...well either Newton is a fraud or the Big Bang is a hoax because both can't be correct!

Now marry that thought with the ever expanding Big Bang beginning and the Newton's concept of a Universe shrinking which totally contradicts the reality that Hubble found to be true that there is a Universe out there of which we are part of that is exploding in expansion in spite of Newton's $F = G \dfrac{M_1 M_2}{r^2}$ applying, or so we have the Brainy Bunch tell us. To the world they declare openly that Newton's contracting Universe and Hubble's expanding is the same thing and we must wait for the Universe to admit being incorrect and start to employ Newton's contracting because even the Universe has to be wrong since Newton can never be wrong.

They put the blame on the Universe for being incorrect in expanding and for the Universe not to contradict as Newton said it does. It is the Universe that is going the wrong way and the Universe being the incorrect party because the Universe expands while Newton stated emphatically the Universe should contract according to $F = G \dfrac{M_1 M_2}{r^2}$. Never could Newton be incorrect and therefore the Universe stands to be corrected by the Critical Density theory because they are looking for mistakes in the Universe and wait to find out when the Universe will start to comply with Newton and start shrinking because the Universe has to stop this ridiculous expanding since Newton said the Universe is contracting.

Since Newton just can't be wrong, therefore the blame of such silly contradicting of Newton has to be found at the door of the Universe. This blame game and detecting how far and why the Universe went wrong in disobeying Newton they named the Critical Density Theory and this Critical Density Theory is the biggest scam and covering of fraud ever invented by any group of persons any time during the history of man. If the Big Bang is true (and it is true), then Newton just doesn't fit! In my book I show how this led to the biggest criminal cover up man has ever devised and was initiated by a person called Albert Einstein. The entire philosophy behind the Critical Density Theory is a scam and is even more ridiculous than what the rest of Newton is. When and if you purchase Newton's fraud you are about to read how far Newtonians will employ their criminal cover up to form a blanket of deception, with one goal and that is to save the arse of Newton and cover Newton's fraud!

I have never found one educated in person physics that would agree or disagree with me in the science of physics. It is your duty as the next generation to test the previous, to ask your Educated Masters to please explain Newton about the flaws that is so obviously apparent. Get them to explain when there is an expanding Universe with Newton advocating contraction, and don't let them off the hook with the Critical Density fraud they created as a diversion. Wise up and get informed instead of allowing those in charge of physics to brainwash you into submitting to their methodical mind bending.

One very simple example, which I mention now at this point, but I do not elaborate on this matter any other place in the book since in this book I wish to limit space, used, is mentioning the gravitational constant. If any one wished to bring in an explanation by employing the gravitational constant also introduced in the Newtonian formula $F = G \dfrac{M_1 M_2}{r^2}$ then using this gravitational constant is one of the ultimate bogus ploys Academics use to confuse the public.

Newton first envisaged the idea that it is mass standing in relation to mass that is destroying the radius found between the two objects forming gravity as presented by the formula $F = \dfrac{r^2}{M_1 M_2}$ but subsequently the notion as well as the formula used changed to $F \propto \dfrac{M_1 M_2}{r^2}$. To get Newton's miscalculation $F = G \dfrac{M_1 M_2}{r^2}$ to work with some dignifying crookedness', they devised a constant of sorts going by the title as the gravitational constant and it is this constant holding the symbol **G** in $F = G \dfrac{M_1 M_2}{r^2}$ It is put in place as being the same as all the gravity but is apparently that gravity that fills the space between the Earth and the Moon. Now comes the Newtonian part… This same space filling ingredient called the gravitational constant, which holds a measured value of 6.67×10^{-11} where it is using this value while it is playing its part in filling all the space we find between the Sun and the Earth as well as the Sun and Pluto and everywhere there is space in outer the gravitational constant is the space-filler to have in that space being filled. If you think of space then we have such space filled with a gravitational constant at a value of 6.67×10^{-11}. This was the case in the days when it was accepted that ether was filling the space the gravitational constant filled and therefore ether might have had the value of 6.67×10^{-11}. Then after finding no evidence of ether, the ether that was not filling the gravitational constant was miraculously and by a stroke of Newtonian magic removed and replaced with…nothing…yes, nothing is now filling the space ether filled before they realised ether was not filling the space but the marvellous part is that nothing that the replaced ether took from the ether that is not there the value given to the gravitational constant and now while space is filled with nothing it still holds the measured value of 6.67×10^{-11}.

Newtonians are most adamant the Universe outside material is filled with nothing and nothing form outer space. Let's quickly ponder on this for a second or two and find out how much of this concept is palatable. They are then saying there is a long line of nothing standing the one after the other where one nothing is following the nothing in front while the nothing in front is leading the nothing behind. The nothing is lines forming rings here every ring ends at a point that the nothing that form the line of nothing connect linking in a chain of nothing from the Sun all the way to Pluto and even far beyond, in fact as far as the mind can take nothing and then nothing links in a line even further.

This line of nothing linking in a chain has a measured value that consists of the gravitational constant and each one of any of the nothing we mention has the value of 6.67×10^{-11}. Well even Red Riding Hood is going to sound more believable than does the most intellectual minds in mathematics…and from where does this first thought originate that the universe comprises of nothing…from Newton of course when Newton said that the spin of an object cancels the space in which the object spins or in mathematical terms

Newton's formula $F = G \dfrac{M_1 M_2}{r^2}$ is suppose to hold the definition and is used by those in teaching positions to explain the Universe. To most, it might be incomprehensible to think that those in science are wrong, but more alarming must be the idea that they are wrong while they knowingly persist to apply what is evidently wrong as an example in which they portray the Universe and moreover would their actions be wrong when they continue to repeat such an evident mistake for the sake of not admitting their mistake. In doing so, that would be completely wrong. If everything is in contraction then by now some places should already contracted large areas of space leaving gaps at other places where cosmic holes should by now be in place. If contraction brings about the gain of space in some parts of the cosmos as Newtonians say, then there has to be a part of the Universe that is losing space and by losing space, a shortage of space must bring holes in space to appear where space is reducing!

The gain of one part must bring about the loss of another part. In all of this that I have mentioned thus far in this letter, it does not even form a drop in the ocean compared to the incorrectness I present on science in my books about science. One should think that a book that challenges the dogma of Mainstream science and brings a new view or if only then just another view on science has to be commercially viable and should have some sort of selling potential. One should imagine that there has to be some publisher that recognises the potential and would have the courage if not the business sense to put such a book in the bookshops. But I found that thinking that way is pure daydreaming. A book such as the book I have to offer that takes on Newton and starts to strangle Newtonians has more scope to become controversy than does the Da Vinci code because it affects a wider audience.

Since 1977 I was convinced that there was something amiss in the approach science took on the matter of gravity. In the work I presented I based my theory on the discipline Johannes Kepler introduced. But the more I pursued my goal of forming another gravity concept other than supported by Newton's view that $F = G \dfrac{M_1 M_2}{r^2}$, the more I had to confront the thinking of Newtonian inspired culture when approaching academics. I was never in doubt that Newton was ultimately wrong and that the Newtonian dogma of gravity was incorrect, but never wished to directly attack Newton as a scientist. But in the end I had to change my approach of being polite because it was clear to me that academic culture would not change and see the logical arguments. In September / October 2005 I wrote letters to nine academics heads of departments at Universities in South Africa that was involved in cosmology. Each one I wrote a letter to was heading a cosmology department at a South African University and in the letter I informed them that I intended to show why Newton was incorrect and therefore what mind games they as physics academics and tutors were playing by protecting Newton's dogma and Newtonian religiosity, I was going to uncover and make public the fraud Newton committed and what the extend of the fraud was that they were intentionally covering up. I knew before I wrote the letters that no one is going to take me serious but then there is a price to pay for every mistake one makes and they were blind to what I said when I said I was going to expose them. Even at this point they do not see that they are intentionally covering up Newton's criminal behaviour and hat they are committing intentional fraud by covering up Newtonian misconceptions. They are so involved in a culture of crime it is not possible for them to separate criminal fiction from factual reality. The covering up syndrome and culture prevailing in physics will prevent them from having any negative work on Newton or about Newton to be published.

By merely putting gravity in the Universe that is acting as a mysterious FORCE that is pulling towards a common point in an allocated general centre is rather avoiding the question with simplicity because the question about how and why remains unanswered. Not knowing the answer will leave you empty and unfulfilled because of being a student and not knowing is the same as suicide on a mental level. Ask yourself the following: If gravity pulls towards a centre and gravity holds the Universe attached the question arising from that simplistic answer is then … where is the centre of the universe?

Should and if you decide to read my letter addressed to students it will bring along a new perception about Kepler. Science sees to it that Kepler stays the least appreciated Cosmologist where as in truth Kepler proved gravity, proved singularity, proved space-time, proved the Big Bang, proved every dynamic most of the wise persons afterwards thought about. Yet, no one gave Kepler any recognition up to now because science denies Kepler his limelight.

By no one being in a position to confront those forming the establishment it means everyone embraces the establishment and their opinion, you as the students in their power, you are the ones which give the establishment grounds to allure you into being sheepish and follow them without asking even simple questions. It is your attitude that allows them to see you as just another stupid mindless, thoughtless creature and just being a senseless student. This attitude you have makes it easy for them to control whatever you think. You supply them with the opinion that they can brainwash you into mindless obedience by applying mind control as they control the information you receive without you having a chance to question or to argue Newton's propaganda. With the absolute draconian control they have on students they forcefully subdue students into accepting these fallacies that I am about to tell you. They will literally brainwash and condition your mind to accept what they never yet were able to prove. If you feel I come across far too strong then put correct values in place

of the symbols in $F = \dfrac{r^2}{M_1 M_2}$ or in $F \, \alpha \, \dfrac{M_1 M_2}{r^2}$ and see according to your own opinion how

totally ridiculous Newton was and physics at present is.

They think that your naivety makes you mindless and that will incapacitate you into their control. They don't want you to ask nosy questions about contradictions existing and they refuse to answer. This process of brainwashing and mind controlling in physics has been in progress for hundreds of years. Just answer how a feather and a large hammer fall equally while mass drive gravity as a force. If you can't…well they can't either! Their task is not to explain but to mislead, since they think you can't think while they think they know how to control you.

As a diversion tactic they formed the Critical Density Theory and spearheading this theory was a man going by the name of Albert Einstein. This came as a result of E. Hubble discovering the Universe was expanding and with the evidence there for everyone to witness, the Newtonian paternity in

physics realised Newton's attraction theory $F = G \dfrac{M_1 M_2}{r^2}$ was in shambles. If Newton was wrong,

then everyone would realise how little the Academics n physics truly understood the principles that drives the cosmos. All funding for further research would dry up. If this applied in medical terms, it would be equal to the medical world discovering that oxygen is spread through the body by blood and that the human body is equipped with a heart. It would be the same as the medical paternity discovering that the human body has a heart and blood is flowing. It is the most basic part of science changing altogether. Then the lot got together and devised a plan to divert the attention of Hubble's discovery away from the fact that it proved Newton (including all those members forming the Newtonian religiosity of physics) knew as little about physics as a cat knows about finances. A plan was introduced to prove Newton correct because the entire Universe was proving Newton wrong with every millimetre that the Universe was expanding. The discovery Hubble brought to science was so accurate they could put a scientific age on the Universe. Instead of never being able to use Newton's

formula $F = G \dfrac{M_1 M_2}{r^2}$ to prove when the Universe will end, they now had the means to calculate

when this lot started. If Newton's formula had any reliability, it would be possible to calculate when the demise would come that would end the solar system after which the Milky Way ending could be calculated. Instead Newton's formula $F = G \dfrac{M_1 M_2}{r^2}$ is as useful for calculation purposes as using an elephant in a mouse trap for bait. But with Hubble being correct, then Newton must be incorrect and Newton could never be incorrect because then the lot of those atheists hiding behind Newton's discovery had to admit they were completely wrong and knew nothing about science. Then they got together and devised a fraud so big, the likeliness of this fraud was never seen before and will never be equalled again. They went about and formed a scheme that could serve as a cover-up to hide behind which they could use as a shield so that no one would discover their lack in understanding of the intricate details driving the cosmos. So the modern Newtonian set out to defraud the world in the same manner as their Master Sir Isaac Newton has done centuries ago by diverting the truth about their Master Sir Isaac Newton

It was clear that the Universe was expanding and that would blow Newton's cover. So they charged a man by the name of Albert Einstein to calculate the mass we could find in the Universe. Trying to establish the measure of mass there is in the universe alone outweighs the expression of insane madness. In other books granting me more luxurious space I explain in more detail just how insane Albert Einstein must have been just to accept this order. I show that Albert Einstein had no grasp for reality, but this book does not allow me the required space to do so. Those forming the paternity of Newtonian physics had this problem that Newton said the cosmos is contracting. When Hubble proved the cosmos is not contracting, Newtonians looked where the cosmos went wrong by not following the Newtonian guidelines he so clearly set the cosmos to follow. It has to contract and not expand. I am not going to go into the more complicated issues, but I will show this much which they bungled.

To all those who feel disgusted by me accusing the greatest name in science that ever lived being Sir Isaac Newton of fraud, please go on and prove me wrong!

They put the task to Einstein to find the value of all the mass there is in the entire Universe to see if the attraction is not in place, then when will the attraction come in place. Remember, with this suggestion they put the blame of Newton not panning out to the door of the cosmos. They don't ask why is Newton not panning out, no, they ask when will Newton pan out. That means they ask when will the Universe correct itself and start doing what Newton said it has to do. The paternity doesn't admit that Newton is not applying, they admit to the Universe that will correct its wrong direction of development sometime in the future and start to do what Newton said it must do. Newton said it is contracting, so it is the Universe that must stop this silly business of expanding, come in line, start to act responsibly and do what Newton said it has to do. The paternity laid the blame of being incorrect at the door of the Universe because it is the Universe that will mend its way, turn back and start to contract as Newton said it must.

$F = G \dfrac{M_1 M_2}{r^2}$ This is the formula Newton used with which Newton proved gravity. This is the formula used to show attraction is taking place.

To find the force of gravity, one has to multiply the mass of the Earth (M_1) with the second mass (M_2) and then divide the distance there is between the objects holding mass (r^2). Using these factors by multiplying (M_1) and (M_2) and dividing with (r^2) should present gravity coming from mass. But those members of the Brainy Bunch clan in physics must know this since they have countless PhD and DSc degrees in mathematics and physics between the lots of them. Therefore with such knowledge and such gloomy evidence their scheming has to be nothing less than a planned attempt to commit fraud…and I accuse them on the following grounds.

Now, convince your mind about my correctness. Do the simple calculations and see whether my statements are true about the formula $F = G \dfrac{M_1 M_2}{r^2}$.

The mass of the (M_1) multiplied by the mass of the (M_2). The effect of the multiplying of the mass as a factor of influence decreases the effect it has when the radius becomes bigger.

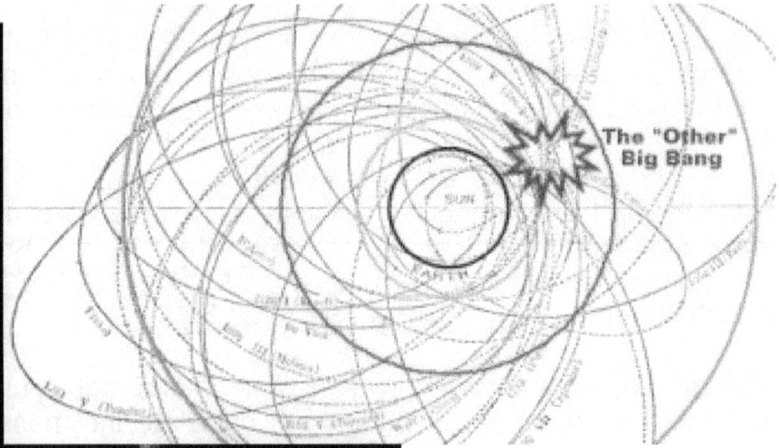

When the radius is very small  r^2  then the effect of mass will be huge.

As the radius become bigger with the cosmos expanding, the radius will reduce the effect that a large

mass and therefore a large gravity would have  r^2 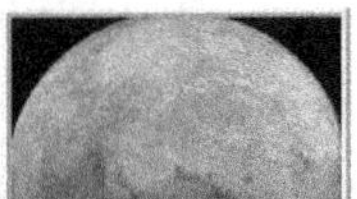 and the mere fact of expanding reduces the attraction by the square thereof.

The bigger that the radius become with the cosmos expanding the, less chance would mass have to

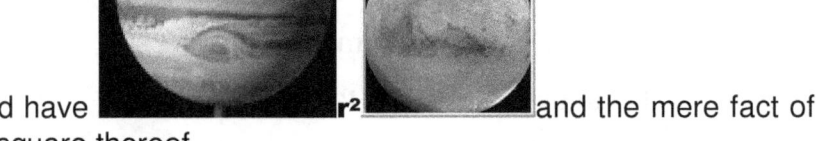

form gravity that would overcome the expanding going on in the Universe  and the fact of

expanding will increase all the while as the expanding improves.

Sir Isaac Newton's formula suggested as $F \ = \ G \ \dfrac{M_1 M_2}{r^2}$ is complete fraud…and no one more substantial than the Universe proves this fact! Unless they prove otherwise, no adding of mass is possible and no removing of mass is possible and with nothing able to change, it makes the Critical

Density Theory the biggest fraud that any group of persons ever plotted. And they got the second biggest Newtonian stooge, namely Albert Einstein to create their diversion. If any one starts to ask questions, then they wait for results that will prove, no not prove Newton correct for Newton is correct. No, they must prove the Critical Density Theory correct! They now must locate dark matter no one can locate. By not being able to detect the dark matter, no one can prove it is there but also no one can prove that it is not there! Isaac Newton can never be incorrect and therefore it must be God and the Universe God created that is wrong and it then is the Universe that has to be corrected!

If there was not enough mass from the start, any adding of mass is not possible because whatever is in the Universe remains the Universe and nothing can remove and nothing can ever add to what is in the Universe.

Mass is an individual factor that is different on anything on which it is applied as a measuring factor. How can something as different as mass that is never constant even on Earth form a constant such as the force of gravity and still be the same in all cases?

Science teaches that a feather and a hammer have different mass while they fall equal in time through an equal distance travelled. All things fall equal in time and distance when subjected to the same environment. If gravity was mass related, then this was not possible, because then objects must fall according to mass. Falling objects bears no evidence of mass playing any part in falling. Any two objects holding different mass, fall equal in time and in distance when sharing similar conditions, which suspends mass altogether as an influencing factor. Galileo proved different mass fall equally under similar conditions. That fact about Galileo, science does embrace, although this strongly contradicts Newton's impressions about mass inflicting gravity. Acknowledging Galileo must make the work of Newton incorrect and also corrupt. On TV we see how all objects, such as cars, humans and bags fall at the same pace, which sets a standard totally against Newtonian mass principles that produce the falling, and proves Newton wrong because mass then does not underwrite gravity in any way or form at all. The formula $F = \dfrac{M_1 M_2}{r^2}$ would suggest mass taking all the responsibility for such falling that takes place. Newtonians declare gravity as the force of gravity F that is equal to gravitational constant G, when it is multiplied by the mass M_1 and the mass M_2 after which then the product of the three factors influencing gravity is divided by the square r^2 distance between mass pulling the mass that destroys the distance between the two objects. If mass pulls mass as Newton said, the Big Bang is not possible, but the Universe is notwithstanding Newton's claims, expanding (growing apart). If the mass destroys the radius separating the objects, then the comet has to collide into the Sun, but it doesn't. If mass forms gravity, every planet must orbit at a different pace, which they do not, as all planets orbit at the same pace around the Sun. Planets don't give the slightest hint that they obey Newton's suggested cosmic laws by implementing mass. The truth is that mass is the resistance of any independent material to deform and to acquire mass the individual object relinquishes independent motion. Mass is the reluctance to deform and integrate into a larger structure and becoming a unit of the larger or holding structure. Mass comes about when the falling of any object stops the motion of the falling. Mass prevents further falling, it does not sustain further falling. Gravity is the moving of the object to the centre of the Earth while falling.

I say gravity is movement while mass is obstructing independent movement, which is what gravity is. Mass is not forming the factor responsible for gravity or movement, but prevents further movement. A body falls by gravity. Mass obstructs further falling, while gravity remains present as a factor that brings the tendency or inclination to move or the attempt to continue moving. Mass hinders movement and therefore mass can't enhance or produce movement or gravity. Mass prevents or blocks gravity. Gravity is the motion that defines the individual identity of any object's structural form by rendering motion while reserving independence in granting free space from other manipulating objects. By saying this, I am awarded the cloak of death by Academics ignoring my correspondence as if I never addressed their mailbox.

This is not the only untruth that the Paternity called Mainstream Science is keeping concealed as a cover up that is wrapped under an airtight blanket of deception. If you sit in class and listen while also experiencing the sinking feeling that the facts you hear are not adding to a total you are comfortable with while you disagree with what is said, then you better read on because this letter addressed to students has it at task to show all that will read this document how much discrepancies academics lay on unsuspecting students that trust Academics with their future and their life.

Do you as students realize the inconsistencies that physic Academics present you with when portraying that which they teach you as being the solemn truth.

Students tell your Professors to stop deceiving and stop trying to control your minds with their fraud. Those Academics tutoring you are telling facts about gravity that has never been proven.
That is mind control.

They wish for you to accept facts on gravity that they hold as the truth. They claim those truths are beyond questioning yet with the least examining those truths they stand by then proves to be totally void of substance because it was never corroborated by one single experiment.

Should you question that mass produces gravity, they will expel you from University by letting you fail your examinations and it was never proven. They will expel you and have you fail tests should you question their authority on the matter of gravity while at the same time they can't for one second bring evidence in support of what they wish you to accept as the unquestionable truth.

That's brainwashing by mind control because if you don't accept their baseless facts as God given truths, they dismiss your academic career.

It is either put up and shut up or be gone. Academics do put mind control to work on unsuspecting students by forcing students never to question the legality of statements they offer as being sound and correct.

What they present as correct, I prove in this very letter addressed to students are openly laughably totally incorrect and by just reading my evidence you will see how feebly easy it is to rubbish it. Take the evidence I am about to share with you and confront them with the fabrication of facts that they present. Go on and challenge those teaching you with the falsified facts as I challenge any one to prove me wrong.

What they maintain is gravity is total incompetent nonsense and can't be corroborated at all but what they can't corroborate because they don't understand I prove to be that which the Universe employs to form gravity. There are four phenomena they dismiss because they have no idea what they are. I studied each one and formed an explaining by implementing Kepler's formula as the Universe gave it to Kepler.
Then by understanding the formula and implementing the content into the four phenomena I am able now to prove what forms the motion we think is gravity and when reading it, then the Universe makes sense. All the questions in these books I managed to answer while they can't … and in the books I answer a lot more questions than what I ask here in the rest of the web site while Science fails to answer any.

Now I am taking my case to the members of the public so that the truth must be brought into the open. I have had the tour they give and then more came my way. I never got around swallowing the mass creating gravity part where science is of the opinion that mass pulls as gravity is… Academics condemned my work and therefore me and for six years where I could not get a publisher to come around and bother to read my work let alone seriously proposing a publishing contract. I had to finally go private with the publishing as all doors shut in my face as soon as the academics read the content of my work because from the nature of my work I take Mainstream science head on and am confrontational on most aspects of astronomy. There does not seem to be any publisher who wants to go head bashing with the establishment of science on official science principles, which I have to do to convey my message in no uncertain language. If you also have doubts about the academics' indisputable correctness, please read on and confront either them or me on everything you read here.

After reading this letter you will have to take sides because you will know the truth.

Then you either become partner in the crime as you cover the truth up or you will be part of the truth with deceit and help me confront them to acknowledge the truth.

Should you think this page is some sort of a prank then answer the following simple question to yourself in utter honesty! If there is a Big Bang with everything moving apart, how does that support Newton's contraction? Tests results received after the Moon landing show the Moon and Earth are moving apart! Yet students learn about mass pulling mass and that puling by mass forces togetherness by contraction.

The entirety of physics rests on this one formula $F = G \dfrac{M_1 M_2}{r^2}$ The questions concerning that which you are studying and that touches every aspect you are academically concerned with, is that if everything is moving apart, how does that support Newton's idea that everything is coming together…and please don't let them fool you with Einstein's Critical Density idea! If there was mass seen or unseen in the Universe and mass generated gravity and gravity does the pulling, then why is the mass not at this moment doing the pulling. What is all that mass of so many supposed stars doing at present while waiting to get to work where it will only later, much later form a force of gravity that then will bring about this pulling of the Universe? What makes the mass slumber in darkness to one day form a pulling force? What has the "darkness" or the fact that we don't see the mass, got to do with the idea that the mass at present is not forming gravity that is forming a pulling force? You are taught that gravity pulls objects to the centre and obviously gravity then has to ultimately pull everything to the centre of the Universe. That is what the Critical density research that Einstein initiated wishes to establish. The idea is that $F = G \dfrac{M_1 M_2}{r^2}$ makes the mass create a force that will destroy the radius and ensure everything is going to come together eventually at one point where the radius then will be no more. If that is the case, then where is that point? If everything is destroying the radius, then it must end at one specific point.

In the classes you attend a physics lecture, has any one confirmed a location where one might find the centre of the Universe to confirm the ultimate destination of $F = G \dfrac{M_1 M_2}{r^2}$? If you wish to apply a Gravitational constant as a calculated factor then it is apparent that one must know to where such gravity is pulling since it then is the gravity that is predominantly keeping everything apart. Then the gravitational constant is what is resisting the collapse of the Universe. If there is a force, then where is the force taking the pulling…if it is a gravitational constant applying through out outer space then where is it having a centre base?

I wrote a book in which I found a means to define gravity. This feat I accomplish and by my effort it was done this for the first time ever. For the first time ever runs further back than since the time Newton introduced gravity. Before I achieved that discovery, I firstly had to find the centre of the Universe because it is there that I could locate gravity. I now am able to show how gravity forms because I have detected the centre of the Universe. But by my effort in finding the location I disrupted everything Academics in physics hold holy and for that I am most unwanted in the presence of the Academics charged with guarding the ethics of physics. In short, I clash head on with Newtonian principles. During my research I discovered abnormalities and inconsistencies about mistakes the Arch fathers in physics must be aware of but is hiding with all their considerable influence. I will come to some of the inconsistencies later on but the discovery also introduces a much better vision about many new aspects that I discovered but in reality was never before realised in science. But these discoveries discard and blacken the Newton reputation totally and therefore the academics dispute my work totally in order to save their Newtonian reputation. The road I took in my search for truth concerning physics was never smooth and the resistance I came across coming from the academic sector was almost unbearable. Academics guarding physics will never allow an outsider to enter their domain and dislodge Newton from being god that is without the intruder paying a heavy price for trying to do so and in this matter I was and still I am seen as being in the role reserved for such an intruder. It is not about my work they detest but it is my rebutting of Newtonian thoughts that they reject! However, such an intrusion allowed me to find so much that I was not supposed to find, which was reserved to all who studied physics. This insider information that is available is only allotted to the most inner circle and this insider information that I share with you. By finding the centre of the Universe enabled me to find a point the Universe is controlled from. In achieving the location of the centre of the Universe I had to step on some very important toes, which made me very unpopular. With my unpopularity rating this high, I never qualified for help and those who would help, found my ideas intolerable whereby I only found rejection instead of help as I tagged along. Because of this insider rejection I had to resort to private publishing because from the nature of my work I take Mainstream science head on and am confrontational on most aspects of astronomy. This is the only road to go if one wishes to lay axe to the root of the insider corruption they are guilty of. In that sense there does not seem to be any publisher who wants to go head bashing with the Physics Custodian establishment of science on official science principles, which I have to do to convey my message in no uncertain language. I argue that if it is the correct practise to use $F = G \dfrac{M_1 M_2}{r^2}$ to calculate gravity, then the radius holding the gravitational constant must lead one to the centre of the Universe. With nobody willing to publish my work as I confront science dogma and principles all the way, I had to go the road alone and fight the battle by my private effort.

This is only one of many points that I make on this one issue and there are so many other issues one may think of those in terms of counting in numbers in many hundreds or even in thousands. If the Sun for instance has mass that is apart from the Earth and the Earth also has mass and there is a gravitational constant in between the Sun's mass and the Earth's mass, we have the radius in that location. It then must be the gravitational constant that fills the space that the radius holds. It is rather obvious that while the radius is filling the vacant space between the Sun and the Earth, it is the only place left where the gravitational constant can hide. To find the centre of the Universe, I had only to find the gravitational constant that holds the centre. Through my venture I discovered one person that knows what gravity is! Newtonians went and filled that space reserved for the gravitational constant having a measured value with nothing! How can nothing have a value of 6.67 X 10 $^{-11}$ while also being filled with nothing as it is nothing filling the nothing of outer space?

If you think scientists know what gravity is, do not be duped that easily because no one in science remotely knows what gravity is...not even Newton knew what gravity is except Kepler... and because of what Kepler introduced, I now know I can prove what gravity is. Gravity is precisely what Kepler said gravity is and only Kepler knew where to find the centre of the Universe because only Kepler knew what gravity is all about.

Try to get an answer from any academic person in physics about where the centre of the Universe is, is like trying to touch the moon.

Newton changed his initial formula that was $F = \dfrac{r^2}{M_1 M_2}$ to $F \ \alpha \ \dfrac{M_1 M_2}{r_2}$ he placed $F = \dfrac{r^2}{M_1 M_2}$ in context to $\left\{ \dfrac{F}{1} = \dfrac{m_1 m_2}{r^2} \right\}$ and by changing the formula by only changing one symbol α the entire outcome of the formula changed without changing anything. Newton saw it fit to replace ▆ with α and the formula was reborn in value while staying the very same. There is an applying rule or law in mathematics that says when one change a formula from $F = \dfrac{r^2}{M_1 M_2}$ to $\left\{ \dfrac{1}{F} = \dfrac{m_1 m_2}{r^2} \right\}$ then

F being F ÷ 1 must also remove a position to become 1 ÷ F making F the fraction value. All those that know even the least about mathematics and of which Newton and his followers not part of knows very well that if any part on the one side changes dynamics from being on top of the dividing line then the very same must apply on the other side. One can't just say that to change a formula

$$F = \frac{r^2}{M_1 M_2} = \left\{ F \; \alpha \; \frac{m_1 m_2}{r^2} \right\}$$ would not translate in ultimately change the outcome of the

formula because the truth about mathematics is that $\left\{ F = \dfrac{r^2}{m_1 m_2} \right\} \neq \left\{ F \; \alpha \; \dfrac{m_1 m_2}{r^2} \right\}$ but when

it changes one then every changes to $\left\{ F = \dfrac{r^2}{m_1 m_2} \right\} = \left\{ \dfrac{1}{F} = \dfrac{m_1 m_2}{r^2} \right\}$. Newton had this idea

that because he was Newton the Great (Cheat) normal rules did not apply and with him being Newton even mathematic laws was below his status. He could replace symbols ▇ with ⍺

$$F = \frac{r^2}{M_1 M_2} = \left\{ F \; \alpha \; \frac{m_1 m_2}{r^2} \right\} = \left\{ \frac{F}{1} = \frac{m_1 m_2}{r^2} \right\}$$ and that will change mathematics

forever. It never dawned on him or his followers that came after him that

$$\left\{ F = \frac{r^2}{m_1 m_2} \right\} \neq \left\{ F \; \alpha \; \frac{m_1 m_2}{r^2} \right\}$$ but the correct application is in

fact $\left\{ F = \dfrac{r^2}{m_1 m_2} \right\} = \left\{ \dfrac{1}{F} = \dfrac{m_1 m_2}{r^2} \right\}$. But then he went much further and cheated the cheated

by introducing $F = \dfrac{r^2}{M_1 M_2} = F = G \dfrac{M_1 M_2}{r^2}$. There was never one Newtonian that even hinted that

the Newtonian could explain how did the initial thought of $F = \dfrac{r^2}{M_1 M_2}$ than mathematically changed

to $\left\{ F \; \alpha \; \dfrac{m_1 m_2}{r^2} \right\}$ which was intended to become $\left\{ \dfrac{F}{1} = \dfrac{m_1 m_2}{r^2} \right\}$ and then with normal,

mathematical principles still applying change this lot to $F = G \dfrac{M_1 M_2}{r^2}$ Furthermore, how could

academics in mathematical physics teach children or students in physics this as the truth! How could

any mathematician explain a process of following logic maintain that $F = \dfrac{r^2}{M_1 M_2} = F = G \dfrac{M_1 M_2}{r^2}$

…explaining it is preposterous.

Let any academic mathematically show how one would go about and use Newton's visionary formula $F = G \dfrac{M_1 M_2}{r^2}$ to calculate the force of gravity by replacing the symbols with the actual values in mass that the symbols should have. Put in the Earth's mass in place where it belongs and put in your mass in place where it should be and then divide that with the distance between your soles and the Earth measured in micro millimetres by the square thereof! If it can't be done, then that is proof of Newton committing fraud when he introduced the formula $F = G \dfrac{M_1 M_2}{r^2}$ being able to calculate the force applying as gravity. Take any formula used in daily physics and show where they use the mass of the Earth as a factor in calculating anything. Never, not once, do any formula used by physics hint that the Earth's mass has any influence on any part of physics when any one calculates factors to determine whatever they wish to determine. If the Earth's mass is never used in any calculation, then the Earth's mass has no part presented as a factor and then the Earth has no mass that influences any aspect of physics. That means the Earth's mass doesn't produce gravity because if it did, the calculating formulae used in physics must use the Earth's mass as a factor in all calculations! Newton cheated to bring in the Earth as a factor that has mass that produces gravity and never does the mass of the Earth contribute to any part in any of the many calculations that form part of physics. The Earth has no mass because the Earth's mass never plays a part in any formula. It is as simple as that! The formula Newton first devised has not even a ring of truth to it. If it is true then show how the

formula reading $F = \dfrac{r^2}{M_1 M_2}$ is used to indicate that this brings about gravity without cheating it to

become $F \; \alpha \; \dfrac{M_1 M_2}{r_2}$ and then committing blatant fraud in changing the formula to able

$F = G \dfrac{M_1 M_2}{r^2}$ while even in this form it still doesn't apply.

The scientific presumption is that gravity is established when one object holding mass is pulling another object having mass and forces the two objects to move toward each other. The Newtonian formula $F = G \dfrac{M_1 M_2}{r^2}$ explains the comet arriving at the Sun, drawn by the mass of the Sun, pulling the mass of the comet as the comet comes closer to the Sun, but then if Newton's $F = G \dfrac{M_1 M_2}{r^2}$ has any validity the comet has to crash into the Sun after arriving. If gravity by mass was pulling the comet towards the Sun in the manner as Newton insisted in the Newtonian formula $F = G \dfrac{M_1 M_2}{r^2}$, then try and get any academic to explain why and how the comet moves away from the Sun and into the black yonder. After reaching the comet, the comet avoids colliding with the Sun as the formula $F = G \dfrac{M_1 M_2}{r^2}$ would suggest and heads into the darkness of outer space. The comet then is moving directly in the opposite direction of what Newton's formula $F = G \dfrac{M_1 M_2}{r^2}$ would have us believe as the comet is not supposed to be pulling away because it is the mass pulling that was in place when the comet was drawn by mass as Newton stated. Does mass then start pushing mass to get the comet floating away from the Sun? Mass establishing gravity by pulling of a force is a gimmick Newton suggested but is unproven and it is nothing less than foolhardy to believe that mass does the pulling of the comet. Try and get those academics in physics to sensibly admit this reality and then in the explaining be sensible by using their Newton formula $F = G \dfrac{M_1 M_2}{r^2}$ as Newton's formula presents the law to show how this going away happens when mass is doing all the pulling at first. Try and get any Newtonian academic to explain this escaping of the comet from the mass of the Sun in the face of mass pulling mass. Some try to use the idea that the momentum drags the comet around the Sun but the mass will pull the comet into the Sun if $F = G \dfrac{M_1 M_2}{r^2}$ applies. Newton never created a detour as the mass pulling mass forms a linking straight line running from the centre of the Sun to the centre of the comet. Newtonians always bring more deceit to cover up Newton's fraud. These questions I address are otherwise never asked by students because students are brainwashed to accept and not think about asking questions. In presenting my work I can and I do answer the questions raised above but my answers do not fit the Newtonian visions of mass doing the pulling and because it contradicts Newton, I am ignored. In my following describing Newtonians is not to moan and grumble but it is to show the means and the manners they use to fight and when using such utter arrogance, despicable high and mighty autocracy with plain bullying tactics and megalomania. They have this attitude that only they are wise enough to think and the rest is mindless dehumanised animals walking on hind legs. If they fought fair and used intelligence it would not be that bad but to use dirty tactics when confronting me by just dismissing my views from a position of having authority is coward ness. By bullying me from holding a position of being able to ignore me and I can do nothing about it doesn't frighten me, it angers me!

Newtonians are most adamant that the Universe outside material is filled with nothing and that nothing forms outer space. Let's quickly ponder on this for a second or two and find out how much of this concept is palatable. They are then saying there is a long line of nothing standing the one after the other where one nothing is following the nothing in front while the nothing in front is leading the nothing behind. The nothing is formed by lines that connects to rings where every ring ends at a point where the nothing that forms the line that connects to nothing and this lot then links in a chain of nothing from the Sun, going all the way to Pluto and even far beyond to Kuiper's belt and even much further and in fact this line holding a continuing connections of endless nothing goes as far as the mind can take nothing and then nothing links in a line even further. I know that you are not going to believe me, but please take my word for this: this is the way that the supposed to be smartest human minds that the world has ever had are thinking and that is the logic that the Newtonian atheists are capable of! ...and everyone considers them to be smart!

Let's review the Newtonian thinking prevailing about what forms outer space. There is the thought amongst the Brainy Bunch of nothing running from the Earth to the Moon and all the surrounding space we call outer space is a nothing linking one bit holding nothing to another bit holding nothing and this combined accumulation consisting of nothing is continuing unabated, while this nothing is forming a chain that all consists of nothing. As far as the mind can go we have nothing forming distances, between planets and the Sun, between stars in the Milky Way, between various galactica, it is one big nothing having something here and there and only the something we call planets and stars are something while all the rest formed by nothing, is filling the space in between some things. Let's take what I said into reconsideration. From where the Earth ends to were the Moon surface starts, the entire space in between is filled with nothing or that is what the Newtonian brilliant minds controlling physics wish to have us believe. The distance between the Earth and the moon is on average 384,400 km of on going and continuing nothing. The distance NASA has to cross to get from the Earth to the Moon is nothing, a total distance of 384,400 km of nothing filling space. When I questioned the validity of this argument of filling outer space with a substance called nothing, and also argued against the idea that the Newtonians actually having a measured value on nothing, because according to their philosophy, nothing can form a distance and according to their wisdom nothing therefore can be measured in physics in one of my books, **A Cosmic Birth Dismissing Nothing**, that book was shot down in flames while I was being accused of incoherency because I question their validity when they are capable of finding it possible to put in the distance that runs between the Sun and any planet a total that comprises the value of nothing. This nothing forms the measured value comprising of outer space and when that distance is filled by nothing, it is then still possible of forming a value that can be measured. To think that nothing can have substance to such an effect that while every micro particle along the way from the Moon to the Earth is formed by nothing going on and going on, and that nothing has the measure to actually form an actual physical distance with substance of having a distance where that distance of nothing has to be crossed to get to the Moon, nothing then forms all of that distance NASA has to cross while it is only containing nothing. By God, if that isn't madness then the Newtonian mind has become so diluted with rubbish, they don't know what is making sense any more!

This goes further than just that. This blob of nothing holds gravity, a force of substance! It is Newton in person that placed nothing in outer space ($\frac{dJ}{dt} = 0$) while at the same time he gave that nothing a gravitational constant value of 6.67 X 10^{-11}. According to Newton, nothing has a measured value that consists of the gravitational constant and each one of any of the nothing we mention has the value of 6.67 X 10^{-11}. This is the mind and thoughts of the most brilliant physicist that ever lived. This is how the father of physics argued when he established the basis for physics. This is the brilliance of the man who has never been proven wrong yet! This is what he said when he said $\frac{dJ}{dt} = 0$ which is that the spin of an object cancels the space in which the object spins or in mathematical terms.

I wish to repeat the following since this forms the motto of the book you are reading: Newton's formula holding and explaining the Universe portrayed as being $F = G\dfrac{M_1 M_2}{r^2}$ is completely wrong.

If everything is in contraction then by now some places should already contracted large areas of space leaving gaps at other places where cosmic holes should by now be in place. If contraction brings about the gain of space in some parts of the cosmos as Newtonians say, then there has to be a part of the Universe that is losing space and by losing space a shortage of space must bring holes in space to appear where space is reducing! By merely putting gravity in the Universe that is acting as a mysterious FORCE that is pulling towards a common point in an allocated general centre is rather avoiding the question with simplicity because the question about how and why remains unanswered. Not knowing the answer will leave you empty and unfulfilled because of being a student and not knowing is the same as suicide on a mental level. Ask yourself the following: If gravity pulls towards a centre and gravity holds the Universe attached the question arising from that simplistic answer is then … where is the centre of the Universe? I know where it is because I follow the path Johannes Kepler showed. It ism precisely where Newton corrupted the work of Johannes Kepler just because Isaac Newton's stupidity rendered him unable to understand the work of a great man, Johannes Kepler.

With the Critical Density shambles the modern Newtonian set out to defraud the world in the same manner as their Master Sir Isaac Newton has done centuries ago. Newton said the cosmos is contracting. When Hubble proved the cosmos is not contracting, Newtonians looked where the cosmos went wrong by not following Newton guide lines he so clearly set the cosmos to follow. It has to contract and not expand. I say to all of the Brainy Bunch, Newtonian disciples that might feel disgusted by me accusing their great Master, going by the name of Sir Isaac Newton and whom they revere to be the scientist bigger than science in itself of fraud, please go on and prove me wrong! Show that $\dfrac{dJ}{dt} = 0$ is correct, as Newton declared. Show that the original formula as $F = \dfrac{r^2}{M_1 M_2}$ is as correct as you all declare! This is the very formula Newton saw applying as he saw an apple falling from the tree. At this point he still had no cosmic aspirations but only tried to formulate what he saw was happening in front of his eyes. He saw an apple and as he witnessed the apple fall he used $F = \dfrac{r^2}{M_1 M_2}$ as the formula with which Newton proved gravity. At first he did not see planets hurtling towards the Sun. The planet idea must have been an inspiration that took possession of his senses much later after he got Universal inspirations about his insight into God's affairs. Now prove gravity by using this formula. Do the following to prove me wrong. Ask your physics professor to explain what it is in the apple that would recognise the mass in the Earth as to find the attraction that would attract the apple to the Earth. If he can't explain the precise detail as to what is pulling the apple to the Earth, I should say it is about time he acquired one of my more informing books where in I explain what happens in using the law of Pythagoras to explain why the apple falls to the ground. Tell that professor to stop being a foolish sod and get wise and see why the apple is falling. Then he can see how little Newton understood and how little he at present knows about physics, because physics is not magic. Physics is that part of Creation that is something logical that we all can understand and explain.

It should be within his and Newton's ability to explain how it is possible to use mass reaching out for mass in order to generate a force and with mass find the force of gravity. Then he has to explain why would one take the mass on both ends and without mass actually touching or making physical contact, multiply the mass of the Earth (M_1) with your personal mass (M_2) and then divide the distance there is between you and the Earth (r^2). What would such multiplying enlist as to create a force except if he would admit he is counting on magic kicking in at some point. Using these factors by multiplying (M_1) and (M_2) and dividing with (r^2) should present gravity coming from mass. But science uses a fixed value to calculate gravity. Without mass being in physical contact with mass, what would establish a magical connection that would draw each other towards one another? If he or she starts to use electromagnetism as an example it is even more proof they have no idea what is going on in physics because I use a formula to prove that gravity and electromagnetism and nuclear power and weak and strong forces are all the same thing but works on different levels of concentrated displacement of heat.

Now, convince your mind about either my correctness about doubting Newton or Newton's absolute unshakeable correctness in the physics Newton presents. Do the simple calculations to find out precisely how immensely strong you are!

They say take the mass of the Earth (M_1). How can something hidden in the mass of the Earth cross the gap between the Earth and the apple that Newton saw falling while the apple is still hanging on the tree? If he starts to tell you about gravitons, then ask him why they have not yet discovered gravitons although they have named that which they still haven't discovered. The undiscovered but already named gravitons are all part of the cover up of Newton's fraud. Without touching, ask why and how it is stated that the Earth attracts the apple and create the falling of the apple.

Multiply the Earth's mass with the mass of the apple that any scale should indicate (M_2). In what way would this multiplying be productive in creating such a pulling? The Earth and the apple are not in contact at that point, not even in part. They are not touching as to unite any part they may have. The only contact the earth and the apple may experience is in the mind of the onlooker that forces such a contact as to credit Newton with correctness. Let you professor clearly prove what form of contact can be as to reach some bond across space to have the Earth pull the apple. If the wise professor can't, tell him to wise up and start to read my book on the subject about what gravity truly is because at this point your professor wishes to teach you about magic and not physics. Why would one have to multiply the mass of the apple with the mass of the Earth where there is no clear contact between the tow having mass and supposedly is creating gravity?

Let's proceed to a point where we do have mass touching mass and the two objects having mass are establishing gravity. Remember, gravity is the part where the object that is disconnected from the Earth still shows the tendency to move and will move whenever there is a chance to do so. Mass is the pressing of this movement or the tendency to still move even after the Earth blocked further movement by allowing mass to come about. Mass is the blocking effort of the Earth in disallowing the object further movement.

Let's use the formula with which Newton started the lot and find out why he abolished the use of this formula. Let's put $F = \dfrac{r^2}{M_1 M_2}$ to use. After multiplying the two mass factors, then proceed to the following step by dividing the multiplied mass factors with the square of the radius there is between your feet and the Earth (r^2), which should not amount to more than a few billionth of a millimetre.

$$F = \frac{r^2}{M_1 M_2} = F = \frac{(.0000000001)^2 \, m}{120 \, kg \times the \ mass \ of \ the \ Earth} = \text{something so small it can't be valued. This had to}$$

make Newton realise nobody is going to believe him or even consider his proposals seriously about this he conceived as attracting gravity which was clear fraud at best.

That which is devised from the formula should be 9.81 Nm/s^2. Ask your brilliant physics professor to show how one goes from a formula such as $F = \dfrac{r^2}{M_1 M_2}$ to Nm/s^2. How did the time factor become a part of the result of the formula? Then Newton changed the first conception from $F = \dfrac{r^2}{M_1 M_2}$

to $F \ \alpha \ \dfrac{M_1 M_2}{r^2}$ which cheated the entire idea into a joke. With $F = \dfrac{r^2}{M_1 M_2}$ it is the mass controlling the distance and that is true. When $F = \dfrac{r^2}{M_1 M_2}$ did not pan out, Newton should have realised he was on the wrong track, but his cheating mind thought a way out. He did not argue the formula in detail. In the scenario $F = \dfrac{r^2}{M_1 M_2}$ we have mass controlling the outcome but when one uses $F \; \alpha \; \dfrac{M_1 M_2}{r^2}$, it is the distance that is in control of the mass. Let me put it this way:

Mass 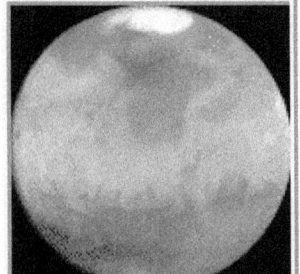 A Large distance reduces  Mass

a small distance puts a seemingly large mass in control.

This is the way we view objects. The further away an object is, the smaller it seems to be and the closer an object is, the larger the object seems to be. That is not the way gravity works since Pluto rotates around the Sun travelling at more or less the very same tempo as Mercury would. That is the reason why a year on Pluto 248.5 x 365.25 = 90764.625 Earth days while Mercury is 88 Earth days.

With the answer of gravity being 9.81 Nm/s^2 there is something very wrong with the use of $F \; \alpha \; \dfrac{M_1 M_2}{r^2}$ because the formula does not deliver time as a factor. The answer given by $F \; \alpha \; \dfrac{M_1 M_2}{r^2}$ has to $F \; \alpha \; \dfrac{kg \times kg}{m^2}$ be kg^2/ meter^2 which might say something convincing about nothing, but it still says nothing convincing at all. Then changing what is rudimentary incorrect to become even more incorrect was Newton's prime ability.

From $F = \dfrac{r^2}{M_1 M_2}$ to $F \; \alpha \; \dfrac{M_1 M_2}{r^2}$ and ending at $F = G \dfrac{M_1 M_2}{r^2}$ helped to change nothing of value very much. Time as a factor remained absent and therefore this formula $F = G \dfrac{M_1 M_2}{r^2}$ is not responsible for calculating this 9.81 Nm/s^2 and in the absence of responsibility $F = G \dfrac{M_1 M_2}{r^2}$ can't take blame for gravity being 9.81 Nm/s^2.

That means that the incorrectness has to be one either of two possibilities presented:
The measured value of gravity is not 9.81 Nm/s^2 as science uses it, or

Sir Isaac Newton's formula suggested as $F = \dfrac{r^2}{M_1 M_2}$ is complete fraud... because using

$F = G \dfrac{M_1 M_2}{r^2}$ just can't produce the small g being 9.81 Nm/s^2.

By using $F = \dfrac{r^2}{M_1 M_2}$ you will float from the Earth and into space going to every Galactica there is

while using $F = G\dfrac{M_1 M_2}{r^2}$ it will crush you into an atomic blob of liquid bloody jelly.

Now which is it…you can decide… or is it neither?

The force of gravity that the world of physics uses to do measurements is 9.81 Nm/s². If the answer you have in calculating your force of gravity is not 9.81 Nm/s², then it is either this measuring value of gravity that is wrong or it is Newton's $F = \dfrac{r^2}{M_1 M_2}$ that is wrong, because by the calculation you did, the calculated answer you got could not possibly have deliver a measured value of 9.81 Nm/s². After all, science maintains it is the pulling of mass that delivers the force of gravity! If by using the factors of mass and the radius does not accumulate to 9.81 Nm/s², then how can mass deliver gravity?

To teach students that $F = \dfrac{r^2}{M_1 M_2}$ can be equal to use as $F = G\dfrac{M_1 M_2}{r^2}$ and this is the prelude to forming the value of 9.81 Nm/s². It is a methodical process of brainwashing students to accept and to believe that rather than to formulate physics by producing proven facts. They maintain $F = G\dfrac{M_1 M_2}{r^2}$ is used to produce the measuring 9.81 Nm/s² and therefore forms that basis formula in determining gravity, while knowing very well it is not totalling gravity at 9.81 Nm/s², then doing that to students while enforcing a thinking pattern in the minds of a student is committing brainwashing because by forcing examinations on students, expecting them to confirm the falsified statements used that the tutors present as correct is brainwashing, a way of enforcing mind control and it is manipulating the thinking process of students.

If you can't prove that my manner of thinking is incorrect and you keep surmising that science is correct, then recalculate the formula or start believing that you are dealing with the biggest fraud ever contemplated by any group of men.

Gravity is a constant of 9.81 Nm/s². This is used in all cases of scientific calculations. Using $F = G\dfrac{M_1 M_2}{r^2}$ allows for an immeasurable range of variations.

Please use your intellect to explain the following mathematical expressions as Newton suggested that the principles of the different formulas applied.

The Academics in science holding positions in Astrophysics and Physics teach students to accept that Newton correctly surmised that physics are fundamentally based on a formula that uses

$$F = G\dfrac{M_1 M_2}{r^2}$$

attraction such as the formula Newton introduced   would suggest.

Your mass is what…let's say **100 kg.**

The Earth's mass is **5.974 x 10²⁴ kg**

The radius between your feet and the earth is a few billionths of a micron because you are standing on the earth. Let's say the distance between your feet and the earth is **10⁻¹² meters**

$$F = G\dfrac{M_1 M_2}{r^2}$$

Take Newton's formula as   and then substitute the values given with the

symbols and see what the answer is.

It is going to unleash a force so big you will not even have atoms left in your body!

When the biggest fraudster that ever lived namely Isaac Newton saw this was completely incompatible with reality and incomprehensible with the truth, he went and devised a new scam. Sir Isaac Newton cheated the laws of mathematics by corrupting mathematical principles. Isaac Newton thought he was superb enough to alter maths. He thought he was so bright that he could change the totally ridiculous formula of $F = \dfrac{r^2}{M_1 M_2}$ to form the formula $F \, \alpha \, \dfrac{M_1 M_2}{r^2}$ by only changing

$$=$$

$$\alpha$$

to then be   . Come on, you are good with mathematics... use all the values you used in the formula $F = \dfrac{r^2}{M_1 M_2}$ as it would apply when expressed as $F \, \alpha \, \dfrac{M_1 M_2}{r^2}$ and find out how this affects the value of F the Force.

I put this challenge to you: in your next maths class go and change any mathematical equation from reading like follows $F = \dfrac{r^2}{M_1 M_2}$ then transform it to directly go through a metamorphoses and read as this $F \, \alpha \, \dfrac{M_1 M_2}{r^2}$ and see what your maths teacher does to your paper. Doing the change is completely mathematically unlawful because the answer you get can't be the same in a million years. This also did not work because every aspect of the formula is incorrect but what does Newtonians do to solve any problem? They cheat their way out of the problem by extending Newton's fraud even further.

Suddenly when Newton's conspiracy doesn't prove true, he and his criminal gang go further to conspire more. If the formula $F \, \alpha \, \dfrac{M_1 M_2}{r^2}$ still doesn't work, they will cheat some more. They

$$\alpha$$

$$=$$

will firstly again change the symbols used from   to then be   as if there is no difference in the use thereof. This is not where they stop cheating. Miraculously, they get a value from the blue and the formula then still remains the same although G as the gravitational; constant is added.

$F \, \alpha \, \dfrac{M_1 M_2}{r^2}$ becomes $F = G \dfrac{M_1 M_2}{r^2}$. We now have a totally new formula replacing the old formula but through the miracle of the magical FORCE called gravity (and Newton's unscrupulous cheating the old formula $F \, \alpha \, \dfrac{M_1 M_2}{r^2}$ that then became the newly invented formula of yet another version $F = G \dfrac{M_1 M_2}{r^2}$ and if you never believed in magic....think again because this proves that physics is founded on unexplained forces that provides magic. The magic the force of

$$\alpha \quad = \quad G$$

back to   and adding   while

gravity produced changing by adding

declaring it still remains the same value. If you think that this is where Newton's surprising magic ends, then you are in for a shock. Wake up and smell fraud for you are going to see how far the deception can go.

They are of the opinion that Newton's vision of his gravitational $F = G\dfrac{M_1 M_2}{r^2}$ attraction applies

as the mass of the earth pulls the mass of a body down to the ground. Then when they start using a

formula that is useful they change the format $F = mv$, which they use considerably! Then by doing that, those academics use their split tongue they inherited from the snake to preach because they are as false as the snake. The main question they never address is what mathematical

road did Newton and his gang of crooks follow to get from $F = G\dfrac{M_1 M_2}{r^2}$ to the formula that is in

daily use as $F = mv$. You can go to any mathematical lecturer and ask whether the route I follow is correct but this is the way it should be done mathematically.

If $F = G\dfrac{M_1 M_2}{r^2}$ and $F = mv$ then we have established that

$F = mv = G\dfrac{M_1 M_2}{r^2}$ because both equations are equal to the same value as F

symbolises. However we have $F = \dfrac{mv}{1}$ which then shows that

$F = \dfrac{mv}{1}$ Then $\dfrac{F}{m} = \dfrac{v}{1}$ and also with $F = G\dfrac{M_1 M_2}{r^2}$ we have the

situation where $\dfrac{F}{M_1} = G\dfrac{M_2}{r^2}$ If $\dfrac{F}{M_1} = G\dfrac{M_2}{r^2}$ and also $\dfrac{F}{m} = \dfrac{v}{1}$ then with

$\dfrac{F}{m} = G\dfrac{M_2}{r^2}$ we have $\dfrac{F}{m} = \dfrac{v}{1} = G\dfrac{M_2}{r^2}$, which means the force divided by the mass of the object is equal

$v = G\dfrac{M_2}{r^2}$ And also we have $\dfrac{F}{m} = G\dfrac{M_2}{r^2}$

In the case where $v = G\dfrac{M_2}{r^2}$ please let those academics show how on earth g can be 9.81 and

They say that $F = \dfrac{r^2}{M_1 M_2}$ becomes $F\,\alpha\,\dfrac{M_1 M_2}{r^2}$ and then also become $F = G\dfrac{M_1 M_2}{r^2}$ which

then becomes $F = \dfrac{mv}{1}$. Let any one of them prove this. Let any one of them explain this as a

mathematical principle. This is all mathematical formulated expressions and has to be proven accordingly. This is not linguistic suggestions but is used in terms of mathematical accountability! This is palpably false

Mass is an individual factor that is different on anything on which it is applied as a measuring factor. How can something as different as mass that is never constant even on Earth form a constant such as the force of gravity and still be the same in all cases?

Sir Isaac Newton's says that $a^3 = T^2$. I have to believe Sir Isaac Newton when it is said that three dimensions are equal to two dimensions or in mathematical terms that $a^3 = T^2$ on no more grounds than that Sir Isaac Newton said so and without having any other proof to back the statement. Remember, Kepler never said $a^3 = T^2$, that is the part coming from the fantasy of Sir Isaac Newton. Kepler said $a^3 = kT^2$ which places three dimensions on one side holding three dimensions equal on the other side of the equation. There is a^3 on the one side of $=$ and then there is kT^2, which is $k^1 \times T^2$ which is $k \times T^2(^{1+2=3})$ and that makes $a^3 = kT^2$ having three dimensions on the one side being equal to three dimensions on the other side. There is no way in heaven or hell that one can have the third power being equal to the second power or have a cube that is equal to a square, even if you are Sir Isaac Newton. There is no one on Earth that will tell me that $10^3 = 10^2$. There is a case that $10^3 = 10^2 \times 10$ or that $2^3 = 2^2 \times 2$ but never can it be that $2^3 = 2^2$. Not even when Sir Isaac Newton is doing the saying so. If one says that in the event where $a^3 = kT^2$ one may assume that $a^3 = a \times a^2$ or $k^3 = k \times k^2$ or even that using $T^3 = T \times T^2$ will also bring equality but never can $a^3 = T^2$…and then there are academics who try to convince me that $a^3 = T^2$ because Sir Isaac Newton was of the opinion that $a^3 = T^2$ and furthermore they expect me to also believe that it is true that Sir Isaac Newton has never been wrong on any suggestion and because no one could ever find Sir Isaac Newton to be wrong, I have to accept that $a^3 = T^2$ and take it as the absolute truth without questioning this abnormality!

The one image is a cube with three sides. The other totally different image is a square having two sides. Sir Isaac Newton said the two are equal while they can never be equal since they are one dimension apart. Sir Isaac Newton convinced so many generations of idiots considered as being the wise amongst the wise and fooled those to the point where these stooges are willing to believe they are wise enough to believe that a cube is equal to a square and only on the ground that Sir Isaac Newton said so.

Sir Isaac Newton proposed and moreover convinced the world of science, and this includes every one and all members that should be the most intellectual bunch living on Earth in human form, that they and the entire world should accept that the inexplicable $a^3 = T^2$ is correct and that the biggest trick in fraud can be played on a bunch of fools all willing to be stupid enough to pretend they are clever enough to see that $a^3 = T^2$ and they are so stupid they pretend to be so clever that they will accept that $a^3 = T^2$ which when translated in words means that two dimensions are equal to three dimensions. This is the same as stating that a person's reflection coming back from the mirror is the same as the person filling reality while standing and looking at his image in the mirror. In this group hosting the most advanced minds man can produce, there are a big enough bunch of zombies pretending to be mentally superior while being big enough idiots who are foolish enough not to think and not to ask questions but be small minded to the point that they will accept that a cube is equal to a square $a^3 = T^2$ just simply going on the say so of Sir Isaac Newton.

When Sir Isaac Newton says $a^3 = T^2$ that does not prove that $a^3 = T^2$. It only proves Sir Isaac Newton was the worlds biggest and best silver tongue devil who cheated an entire Earth load of scientists for almost four hundred years. He fooled the supposedly wisest humans we all think there can be to pretend to be wise so that they can hide their stupidity while they only focus on their stupidity by not questioning the validity of $a^3 = T^2$. You bring me one other con artist and fraudster that can manage that. It takes some doing to fool so many people for so long and leave all those fooled feeling good about themselves in that they are fooled. Sir Isaac Newton was the biggest con artist ever to live and never again will the world experience an equal to Sir Isaac Newton. It is no small wonder that science is infested with atheism because science upholds disdainful lies based on mediocre understanding about truth applying as a reality and crooked science! Newton is all lies and shambles and reading this book will prove that.

If science cannot prove God's existence, it is not God that does not exist, but it is science failing and therefore it is then that specific view about science that should be re-examined since it is the view on science that is proving as being incorrect. This fact is what the so very brilliant and intellectually mindful Newtonian atheist should remember when they fail in their science altogether. That their science fails altogether and that failing it does in all its splendour, is a fact I am delighted to prove! The fact is Newton's views were never tested and that the Newtonian views on science were never challenged before and because of that Newton principles never withstood diligent scrutiny before. When Sir Isaac Newton is investigated even in the flimsiest of manners, well accepted facts seem to become very suspect, to say the least. This becomes evident when concluding all the facts this book presents. Now, in this book, for the first time Newton is tested and such testing is the proof you gain by reading that which I uncover. What I bring into the open are unseen facts, which I present you with as I take you on a tour through an avenue of facts I introduce in this work. The lack there is in sensibility concerning Sir Isaac Newton's principles, this book proves. The theories of Sir Isaac Newton require proof, which was never given while God never needs proof and that is what science constantly seeks. When science perpetually ignored my concerned calling on them because (I suppose) they were finding my concerns wanting, in my final letter to them I promised them never to contact them personally again by any and by all means. I also promised them a fight. This is the fight I promised. I was not worth noticing so I was ignored. I now am calling on the public, as I am ignoring their reputations.

All prospective, intending and otherwise possible potential readers of anyone of the two books called An Open Letter On Gravity are hereby seriously advised to read Part 1 of An Open Letter On Gravity first before advancing onto Part 2 of An Open Letter On Gravity, and only then afterwards and after completed reading Part 1 of An Open Letter On Gravity, then the reader should advance by reading Part 2 of An Open Letter On Gravity because by first reading Part 2 of An Open Letter On Gravity the answers might seem to remain questionable but when reading Part 1 of An Open Letter On Gravity first the questions will become answerable. If it is said by using easier language to explain the expression: then by not knowing the questions, the answers are not well defined but knowing the questions, the answers become rather simplified as it is self explained and much better understood in explaining.

Still my challenge is and remains there where I challenge everyone and all persons notwithstanding title or position to show me how they can maintain with clear and lily white conscience that Sir Isaac Newton did not corrupt science in all aspects by committing fraud in mathematics and do so after reading the two titles An Open Letter On Gravity Part 1 and Part 2

AN OPEN LETTER ON GRAVITY Part 1+2 disputes the correctness of the formula $F = \dfrac{r^2}{M_1 M_2}$

Using the formula above as Newton did does not imply a suggestion or carry an idea across as a thought but must be seen to be acting as confirmation about a fact because one cannot suggest anything mathematically, one can only confirm a fact mathematically. There is no mere suggesting of any possible movement in a specific direction of any suspected behaviour by an object moving from and to a point as suggested but this is saying that the gravity of the Earth measured in mass at it's totality is colluding with the falling body's measured in mass as the two factor's diminish the radius from both ends. This is used to back up a fact!

This Is a Book That Is Not Afraid To Show How The Paternity   Newtonian Science in Physics Openly Cheats To Cover Their Oversight In an All Out Effort to Hide Newton's Misjudgement

Newtonians say the force F of gravity is 9.81Nm/s^2
Newtonians also say the force F of gravity is

$$F = \frac{r^2}{M_1 M_2}$$

Then in terns of mathematical principles

Newtonians say $F = \dfrac{r^2}{M_1 M_2}$ is 9.81Nm/s^2 where both are equal to the force **F** of gravity

$F = \dfrac{r^2}{M_1 M_2}$ **= 9.81Nm/s^2. That is the way that proving with mathematics is done and what does it prove? SIR ISAAC NEWTON'S FRAUD**

Science teaches that a feather and a hammer have different mass while they fall equal in time through an equal distance travelled. All things fall equal in time and distance when subject to the same environment. If gravity was mass related, then this was not possible, because then objects must fall according to mass. Falling objects bears no evidence of mass playing any part in falling. Any two objects holding different mass fall equal in time and in distance when sharing similar conditions, which suspends mass altogether as an influencing factor. Galileo proved different mass fall equally under similar conditions. That fact about Galileo, science does embrace, although this strongly contradicts Newton's impressions about mass inflicting gravity. Acknowledging Galileo must make the work of Newton incorrect and also corrupt. On TV we see how all objects, such as cars, humans and bags fall at the same pace, which sets a standard totally against Newtonian mass principles that produce the falling, and proves Newton wrong because mass then does not underwrite gravity in any

way or form at all. The formula $F = G\dfrac{M_1 M_2}{r^2}$ would suggest mass taking all the responsibility for

such falling that takes place. Newtonians declare gravity as the **force** of gravity **F**, that is **=** equal to **gravitational constant G**, when it is multiplied by the **mass M$_1$** and the **mass M$_2$** after which then the product of the three factors influencing gravity is divided by the square **r** distance between **mass pulling the mass** that **destroys** the **distance between** the two **objects**. If mass pulls mass as Newton said, the Big Bang is not possible, but the Universe is notwithstanding Newton's claims, expanding (growing apart). If the mass destroys the radius separating the objects, then the comet has to collide into the Sun, but it doesn't. If mass forms gravity, every planet must orbit at a different pace, which they do not, as all planets orbit at the same pace around the Sun. Planets don't give the slightest hint that they obey Newton's suggested cosmic laws by implementing mass. The truth is that mass is the resistance of any independent material to deform and to acquire mass the individual object relinquishes independent motion. Mass is the reluctance to deform and integrate into a larger structure and becoming a unit of the larger or holding structure. Mass comes about when the falling of any object stops the motion of the falling. Mass prevents further falling, it does not sustain further falling. Gravity is the moving of the object to the centre of the Earth while falling.

I say gravity is movement while mass is obstructing independent movement, which is what gravity is. Mass is not forming the factor responsible for gravity or movement, but prevents further movement. A body falls by gravity. Mass obstructs further falling, while gravity remains present as a factor that brings the tendency or inclination to move or the attempt to continue moving. Mass hinders movement and therefore mass can't enhance or produce movement or gravity. Mass prevents or blocks gravity. Gravity is the motion that defines the individual identity of any object's structural form by rendering motion while reserving independence in granting free space from other manipulating objects. By saying this, I am awarded the cloak of death by Academics ignoring my correspondence as if I never addressed their mailbox.

This is not the only untruth that the Paternity called Mainstream Science is keeping concealed as a cover up that is wrapped under an airtight blanket of deception. If you sit in class and listen while also experiencing the sinking feeling that the facts you hear are not adding to a total you are comfortable with while you disagree with what is said then you better read on because this letter addressed to students has it at task to show all who will read this document how much discrepancies academics lay on unsuspecting students that trust Academics with their future and their life.

Do you as students realize the inconsistencies that physic Academics present you with when portraying that what they teach you as being the solemn truth.

Students, tell your Professors to stop deceiving and stop trying to control your minds with their fraud. Those Academics tutoring you are telling facts about gravity that has never been proven.

That is mind control.

They wish for you to accept facts on gravity that they hold as the truth. They claim those truths are beyond questioning. Yet with the least examining those truths they stand by then prove to be totally void of substance because it was never corroborated by one single experiment.

Should you question that mass produces gravity, they will expel you from University by letting you fail your examinations and it was never proven. They will expel you and have you fail tests should you question their authority on the matter of gravity while at the same time they can't for one second bring evidence in support of what they wish you to accept as the unquestionable truth.

That's brainwashing by mind control because if you don't accept their baseless facts as God given truths, they dismiss your academic career.

It is either put up and shut up or be gone. Academics do put mind control to work on unsuspecting students by forcing students never to question the legality of statements they offer as being sound and correct.

What they present as correct I prove in this very letter addressed to students are openly laughably totally incorrect and by just reading my evidence you will see how feebly easy it is to rubbish it. Take the evidence I am about to share with you and confront them with the fabrication of facts that they present. Go on and challenge those teaching you with the falsified facts as I challenge any one to prove me wrong.

What they maintain is gravity is total incompetent nonsense and can't be corroborated at all but what they can't corroborate because they don't understand I prove to be that which the Universe employs to form gravity. There are four phenomena they dismiss because they have no idea what they are. I studied each one and formed an explanation by implementing Kepler's formula as the Universe gave it to Kepler.

Then by understanding the formula and implementing the content into the four phenomena, I am able now to prove what forms the motion we think is gravity and when reading it, then the Universe makes sense.

All the questions in these books I managed to answer while they can't … and in the books I answer a lot more questions than what I ask there in the rest of the web site while Science fails to answer any doubt that may rise from modern information coming about. Newton is made to be the example of an all knowing God while every speck of information that we obtain from studying the cosmos contradicts Newton totally…and yet do not challenge Newton for should you do, you will get the wrath of God upon you. However, give them the benefit of the doubt by allowing them to prove Newton in Newton's presumption of mass bringing about gravity by attraction.

Tell those persons teaching you to inform you about the precise distance that the Moon and the Earth moved closer during the past forty years. Tell your teacher / lector / professor to use the precise data that is available to everyone. Tell them to use the data gathered since the Moon landing took place in 1969. During the Moon walking at the time, distance measuring equipment was placed on the Moon in order to measure the distance changing between the Earth and the Moon. This data is very precise and readily available on request. Should the person that is placed in such a position as to teach physics, not be able as to provide you with the necessary information, then tell those parties they are uninformed about their subject they teach and should vacate their positions as result of their inadequate abilities and insufficient information they have on the subject they teach.

Those teachers have a choice of three answers and all three will prove that they are dishonest.

If they say they don't know how much the distance of the Earth and the Moon changed in the past forty years, they are dishonest about being knowledgeable about their teachings and are therefore not suited to teach.

If they tell you that the Earth and Moon did move closer together, then they are the deviant crooks that I accuse them to be. Force their hand to prove the claim by supplying substantiating figures.

If they tell you the Earth and Moon grew further apart, then at least they are informed and are honest in their information, but are dishonest in the way that they select the information that they provide

about the working of Newton's gravity. Then you ask them to control that evidence in the light of what they teach you about Newton's concept of $F = G \dfrac{M_1 M_2}{r^2}$ and how it is possible to reconcile the Moon and Earth growing apart with the gravity being an attracting force. If they persist in upholding Newton and his formula of attraction, then you would know who are the dishonest cheats, they or I! Disinformation and the teaching thereof is a method of mind control and brainwashing by manipulating thought patterns and fixing preset ideas. They wish to control what you think by selecting untrue information and then submit you to their rhetorical repetition of unproven facts. That is brainwashing by providing selected information and thus is thought behaviour control and there is just no getting out of the accusation. This is how professors conduct physics and practise teaching methods. Are you sure you like what they are doing to you by falsifying the information they provide and present you with?

That is an act of a criminal mind notwithstanding what motivated the person to commit the crime. Defining a criminal is to have a person that is prepared to place himself or herself in the centre of the Universe as to deprive all others of truth and possession in order to further the needs and wealth of the criminal while the criminal acts by never thinking about the rights of the victim or the harm done by the criminal's action in the matter. Spreading untruths and forcing payment for such actions is most certainly criminal!

You students in physics; go on and confront you Professors about the truth. Insist on the truth for once and all. Mass has nothing to do with gravity but to prevent gravity from being in place. If it is mass pulling the comet as Newtonians declare, then they better inform you being the paying students that keep their wallets filled what is pushing the comet away. The comet is leaving as fast as the comet was coming. Confront them because you have the right to insist not to be lied to about Newton and his shambles and criminal fraud. Go on and ask them to explain these Newtonian inconsistencies and see how they try to carry on with the criminal deception and covering of the truth that has been going on since Newton first thought up this scam.

It is assumed that this is the basis of physics.

$F = G \dfrac{M_1 M_2}{r^2}$ It is not a concept or a suggestion or an idea but it is a formula used as a mathematical expression in terms of a mathematical formula and therefore it is expressed in terms of mathematical accuracy. It has to be put to the test in relation to its mathematical accuracy or Newton is a fake. That is as simple as it is!

This is but a small part of a big picture uncovering the scam Newton came up with and which all the academics are knowingly still participating in… and do read my other books for in those I am about to inform you of more criminality the academics came up with and with which they force feed you. You will see fraud as you never saw fraud before.

Let me tell you something those Newtonians don't want to hear…there is no such a thing as mass and they are stupid in thinking there is mass.

The concept of mass as a usable and tangible factor forms when an object is motionless. When the movement is contained to a position, it does not move any more we think of this situation to be the point of mass. For instance you can't jump on a scale while weighing because the measurement would not be accurate. You can't have a truck rush over a scale at a high speed on a weigh bridge and determine the mass it carries…and don't let them come with this idea of having a difference between mass and weight because that too is so much part of their cheating to let their inconsistencies be hidden under deceit. If mass is measured in weight and weight is measured in weight then both are mass or weight or mass forming weight. You can't have an animal looking like a frog and the other looking like a frog while they tell you the second animal is called a donkey. If the second animal looks like a frog it is a frog and then they are incorrect about the identity of the second frog. If mass uses grams to find measurement and weight uses grams to find measurement, then both use the same to find measurement because both things are the same. There is no way to remain honest and still differentiate between mass and weight because they try again to bullshit you into the dog kennels of being obedient to their systematic mind control they force upon you. There is no such a thing as mass.

Everything is part of movement. Because time moves everything and time is the change of everything in relation to one point, which can be any point, therefore time moves everything in relation to any one single point. However, time is movement and time connects all things and therefore gravity is movement or gravity is time whichever way you wish to look at the issue. Then, what is mass? Mass becomes a measurement factor, a device one may use when one object is standing still in relation to everything else on Earth and when that object does not change in position while being in mass in relation to everything else standing still. The concept is much more complicated than this bit I now mention and I have written an entire seven hundred page A4 manuscript on the matter where I was unable to get any scientist to read it because I begin the manuscript by announcing Newton's incorrectness. Be that as it may, I challenge any person wherever to prove the fact of mass. Mass becomes a unit only when all movement is depending on the movement of the Earth where such movement is in relation to the spin around the Earth's axis as well as spin around the axis of the Sun. Then the movement in mass of mass is represented by the Earth's movement and there is no independent movement of the object that is supposedly having mass.

Everything depends on movement because movement brings time as a factor or a component into space in order to form space-time. Everything is in space-time and nothing is in mass because my mass on Earth is 120kg, but if my mass was measured on the Sun it would be 12 tonnes and if my mass was measured in a Black Hole, I would be billions of trillions of tonnes and less in size than one atom. There is no mass but there is only movement. The object only has mass because the Earth restricts the independent movement to 1^0 or $\frac{dJ}{dt} = 1^0$ and in that Newton was correct but Newton was totally incorrect to say $\frac{dJ}{dt} = 0$. One object standing on Earth would be in a position to show a measurable mass factor, but only as a measurable factor and nothing else when the object is in a position to experience $\frac{dJ}{dt} = 1^0$. This however allows a status to the object and not a condition of the object. No object has mass but could find mass forming a factor when the motion of the Earth excludes the object from forming individual time or if you wish to call it gravity, then do so because time is gravity and it is movement and this allows mass under specific applying conditions such as $\frac{dJ}{dt} = 1^0$. Coming to the conclusion about gravity being motion and mass being the restriction of motion was the easy part. What produced the motion and what prevented the restriction from overcoming the motion was the tough part. Figuring out why everything was on the move and where did the motion stop, that was the part that took some figuring and some explaining. What made gravity move and why does gravity move…the answers are in the four phenomena never yet explained to satisfaction but now turns out to be the cradle of gravity.

However, since these phenomena do not support Newton's mass conceptions and support any evidence of contracting gravity, but disproves Newton's suggestions completely, Newtonians totally fail to understand the workings of these phenomena. It is because these phenomena contradict everything Newton said about gravity contracting, that Newtonians deny the existence of some of the phenomena. It is hardly surprising that Newtonians in some of the cases again even deny the reliability of the phenomena. One should think that where the cosmos shows clearly the existing of such phenomena that contradicts Newton, science would revalue the correctness of Newton. But alas, no…in typical Newtonian fashion the suspsion does not fall on Newton but is reflected back to the cosmos since the cosmos can be at fault, but never can Newton be mistaken. Since the phenomena disprove Newton, therefore it is the cosmos that uses unproven phenomena. Gravity is the movement of space through time and is performed by the following combination of cosmic phenomena:

Gravity is The Roche limit,
   Gravity is The Lagrangian system
      Gravity is The Titius Bode law

Gravity is The Coanda affect

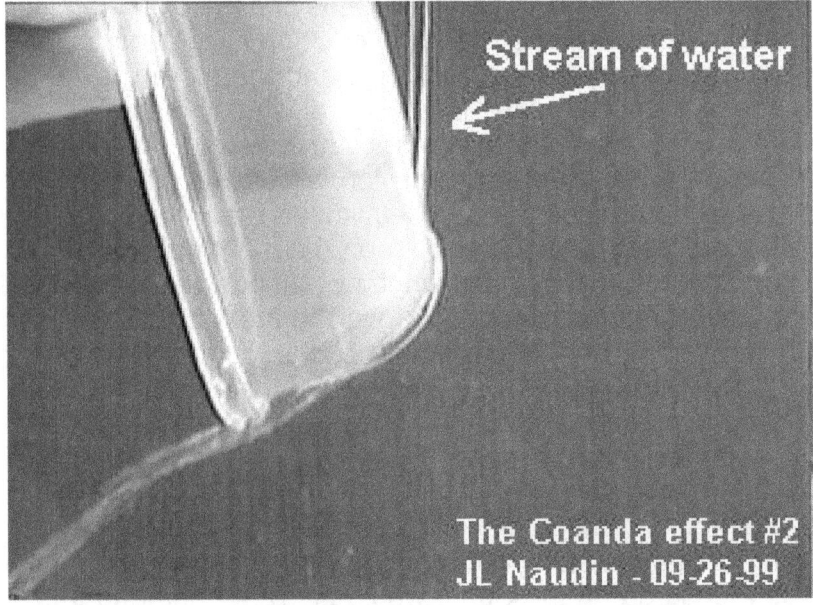

Stream of water

The Coanda effect #2
JL Naudin - 09-26-99

And gravity uses the Roche limit in combination with the Titius Bode law as well as the Lagrangian points to form the Coanda effect from which we can explain what forms the principle in producing the sound barrier. Read the book and find out why this is the case.

In the following books I prove mathematically that the Coanda effect is the principle by which gravity is conducted.

What is the Coanda effect? It is something Newtonians have no grasp of and even less do they know how the principle works.

The Coanda effect is the phenomenon where liquids attach to solids when there is a motion difference between either the liquid spinning or the solid spinning. It is most prevalent where a car starts to run on a layer of water while driving on a wet tar or cement surface.

This is the Coanda effect and is as little understood by Newtonians as general physics is understood be Newtonians.

Ask any professor to explain what this is and how this comes about and he or she will start with a force coming from

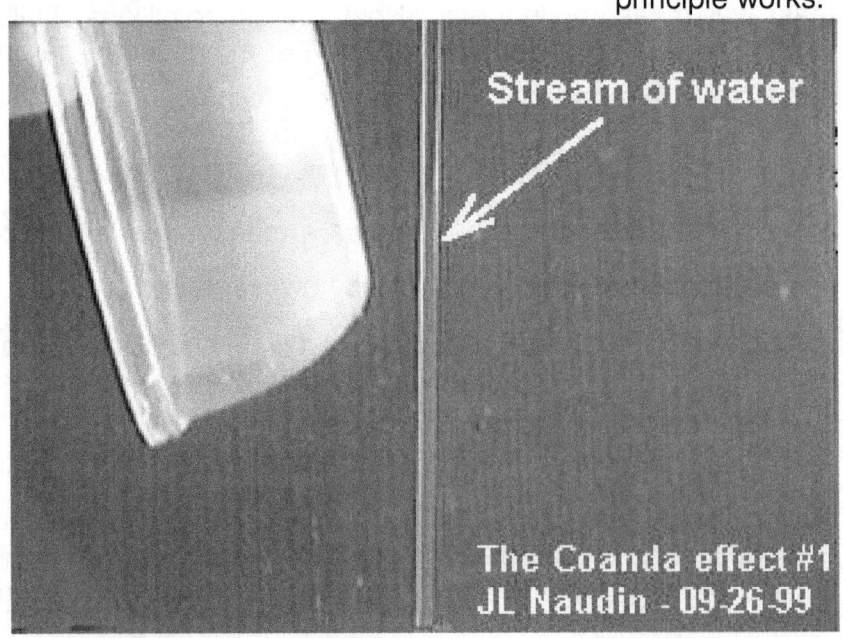

Stream of water

The Coanda effect #1
JL Naudin - 09-26-99

the water and forces coming from the glass and going into some orgy while knowing these forces he adopts for the job clashes with their four official forces they have. Nevertheless, they will dedicate this to the magic of forces contributing to the mysteries of the Universe.

As you can see from the sketch above, Newtonians put the explaining of the Coanda effect squarely on the influences of magical forces, other than the four they already enlist into science. You can take my word for it, although the Newtonians are thought to be so advanced in thinking and notwithstanding the Newtonian's super intellectual explanation about forces in the glass and forces in the water and witches on broomsticks flying around, there are no forces in this concept called the Coanda effect anywhere to be traced.

What you observe in the Coanda effect sounds complicated but it is not when it is correctly explained. The Coanda effect is the differentiation in movement between any round solid structure and any liquid that is in contact with that round object. Should you wish to find true gravity,

The Coanda effect with Air
JL Naudin - 09-26-99

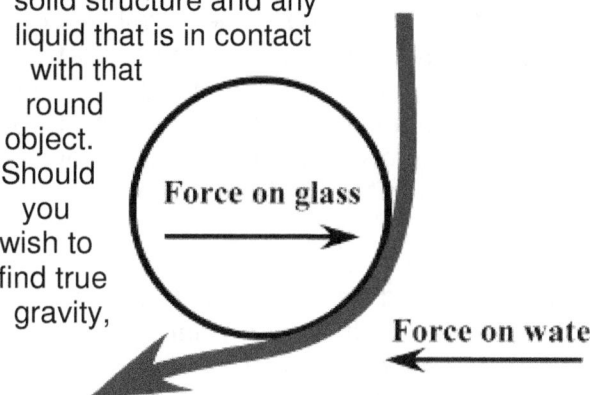

and then replace the solid bowl with the Earth spinning…also replace the liquid water with the atmosphere because engineers calculate the atmosphere just the same way they would regard when dealing with the characteristics of a fluid. Replace the movement of the water with the spin of the Earth and it is most clear that you have the Coanda effect going by another name Newton called gravity. This I explain so simple in other books but to explain and prove the concept takes many pages of simple explaining.

I prove that the Coanda effect is the result of the combining of the following phenomena:
1) The location, the position and the value of **singularity** as a factor forming space-time

2) Finding **space-time** by dissecting Kepler's formula in relation to locating and revaluating singularity.

3) Finding space-time, **proving space-time** and **aligning space-time** with gravity

4) Finding the **working principals** behind and manifesting **of gravity** as a cosmic occurrence.

5) Finding the **Roche limit** and explaining the resulting of a law coming about from singularity.

6) Finding the **Lagrangian system**, how and why that becomes the building form of the Universe.

7) Finding the **Titius Bode law** and I show mathematically how gravity comes about from that.

8) By proving that the **Coanda affect** is producing gravity through reproducing space-time. However since mass plays no part in the explaining of these phenomena and therefore Newton's Mythology does not apply, Newtonians haven't the vaguest of ideas how the cosmic phenomena applies or in fact works.

This is gravity as Kepler proved gravity to be where space a^3 is equal = to movement T^2 in relation to the relevance of k or when mathematically expressed as space-time then it reads as $a^3 = T^2$

That which forms all the information in the entirety of this book cannot serve as an introduction to the information that I introduce, reveal and prove mathematically when reading the four volumes of <u>An Open Letter on Gravity.</u>

Module Four

Welcome to
www.newtonsfraud.com

Yes, astonishingly as it may seem at first glance, but this page refers to none other person, than the person seen as the principle pillar of physics, the person going by the name of **Sir Isaac Newton**. This page will not interest the minds of members holding important titles in mainstream physics, because the information that this web page brings are very familiar to those Super-Educated Master-minds in physics.

Those professors educating students will discourage students to visit this page. However, that will be because students may use the content that such information in these books offered on this page provide, to press, push and demand from their tutors who are the Master-Minds in physics for answers about questions the Brainy Bunch Super-Educated never wish to hear. The books referred in this web page inform students about irregularities concerning facts about the fundamental thought on gravity that those Master-Minds in physics never could address because it shows on what fraud is physics based.

Information obtained from the book/s offered on this web page (and other web pages I mention that you may visit), as well as my other books I offer for sale in **newtonsfruad @lantic.net**, I supply information that is only known to the very top person's in physics. Those Brainy Bunch Super-Educated Masters- In- Deception who have much to lose and little to gain by informing students on these certain matters and concepts in physics. They are the ones benefiting from students remaining in the dark. Those will be desperate to keep this information away from students, because it counteracts their **methodical brainwashin**g with which they teach.

If you are one of those members of society who never thought you would hear the name of an accomplished person such as **Sir Isaac Newton** being associated with **fraud**, **corruption** and **brainwashing**, then these books are specially written to inform you about the truth there is lacking in the correctness about science. Gravity shows no attraction. Planets do not pull and therefore there is no attraction between planets and or the Sun. This is why planets orbit around the Sun and the Moon never will collide with the Earth. The end of the Earth will never come from the Moon colliding with the Earth, or the Earth hitting the Sun in any part of the Future.

Are you under the impression that **everything that Newton ever surmised** about **gravity is proven** beyond doubt? Are you under the impression that everyone in physics knows and understands everything there is to know about gravity and with gravity being the attracting force as Newton surmised, Mainstream Physics is absolutely clued up on the interaction caused by the effect of having gravity as the attracting force?

Are you under the impression that everything Mainstream Physics teach about **anything Newton ever surmised** on the issue of **gravity is proven** and is established as unblemished, sound beyond doubt and is completely proven to withstand all possible forms of doubt? If you are one of those gullible persons, then also you are in for a shock.

The **definition of gravity** is that **gravity is a force of attraction between bodies with mass**…and if gravity is the force of attraction, ask those claiming that Newton's gravity by attraction is the absolute truth, **when will the Sun collide** with **the Earth?** If gravity **is a force of attraction,** then when will the **Moon come crashing onto the Earth?** When will gravity's attraction bring the ultimate dooms day scenario…the destruction beyond Biblical proportions that all dread to think of?

How **valid, reliable truthful and trustworthy** would you say is the Newton **gravitational formula**

$$F = G \frac{M_1 M_2}{r^2}$$ that says all mass pulls all mass closer and also remember that on this formula rests all the fundamentals forming every principle attached to physics! If this formula is as valid and is as reliable as students are brainwashed into accept, then those advocating this idea must be able to measure at what rate is the Moon coming closer to the Earth? Surely those person's with smart enough abilities being able to calculate physics and dabble in rocket science should be able to calculate an approximate date when the ultimate Earth /Moon collision would take place by using the

all so **accurate Newton gravitational formula** $$F = G \frac{M_1 M_2}{r^2}$$ and then permitting the

absolute accuracy of such reliable calculation, we can be schooled in the thought when the end of the Earth will take place?

Does any of the answers to these questions interest you...then please read on for there is much more. If you are tickled by the intrigue that one would find in bringing the Newton gravitational formula

$$F = G \frac{M_1 M_2}{r^2}$$ to any conclusion, then go on reading this web page. If you are a student in

physics at any level or doing any studies founded on the accurateness of physics, then you better read this web page to find conclusive answers to defining questions about physics and the accuracy of what you study! However, if you teach physics, I then accept you will hate and discard all of what I have said this far...and that you will do without bringing any proof about me being wrong even in the least!

Everyone knows that planets orbit around the Sun. Planets circle the Sun which is the same as saying planets orbit the Sun. This circular movement is continuous, unabated, unbroken and is going to be ever lasting. Just by calling the circle motion in terms of what applies in outer space, is a statement nullifying Newton's claim of mass that attracts mass to destroy the radius between planets and that questions the reliability of Newton's dogma. If mass did attract mass, what kept the balance between the planets allowing planets to eternally remain in orbit whereas with Newton's vision applying, then planets would rather then be moving towards the Sun in terms of using a straight line as if they are falling towards the Sun. The mass would pull the mass..., but planets totally contradict this Newtonian principle! This shows fraud and a cover up in terms of Mainstream Physics conducting the methodical brainwashing on students to believe that the falling caused by Newton's gravity attracting is in fact taking place! If mass attracts mass, then what pushes planets back into orbit when they should use a line to move closer and destroy the radius the attraction is committed to remove?

Academics brush this off as one with a small mind trying to ponder on a small issue. Academics brush this off as small insignificant detail best to ignore because this idea carries little significance and contributes in no meaningful way. The idea of proven beyond doubt that Mainstream Physics hold as an unwavering fact being more reliable as anything coming from the Bible is automatically placed at the door of Newton. If normal speech contradicts Newton, then it is Newton's task to prove that Newtonian supposition is absolutely correct in terms of contradicting normal speech on the claims about gravity forcing the attraction in movement there should be between planets which then would prove that the claims Newton made are in fact totally correct. This thought is upheld as being beyond question although this doubting that should arise, Newtonians will deny without bringing proof about fact to show Newton stands corrected. In the past those in Mainstream Physics attacked me personally as they defend their Master whereas they see it as if they prove that the honour and the reliability of their Master Newton with him being infallibly correct. Think of what planets do... and then you think that planets orbit. It is connected to the brain. No one thinks of planets spinning or planets basking in the summer Sun while coming closer. When hearing about planets the first thing that comes to mind is the rotating of planets while circling around the Sun and not spherical orbiting structures resigned to destruct.

However, just using the term orbiting that planets are doing, is in total defiance with Newton! Newton said gravity draws or pulls or moves in the direction…, which would have one understand that the two objects for example the Sun and any of the various planets will be moving directly towards each other. The term pulling does not suggest any circling because no one can be pulling towards and does the pulling while circling around the object. When pulling anything, it must take place while using the shortest line possible. That serves the term pulling. Then the saying goes that planets orbit, indicating they follow a circle. That is not what Newton said. However wrong that may seem, circling is precisely what planets are doing, which means they are not pulling. If they are pulling as Newton claimed, those with the mathematical abilities should be in a position to work out when planets are going to collide? In conversation we speak of the planets orbiting, because that is precisely what they do…they are not attracting…they are endlessly orbiting. If Newton was correct we should be speaking of the planets pulling, attracting, closing in on one another, but talking about pulling in terms of planets circling would be blatantly wrong, that is in terms of and according to the normal spoken word. Never do we refer to the planets pulling the Sun or the Sun pulling the planets, but we speak of seasons coming from orbital positions. Being in orbit has to neutralise the pulling and then cancel the pulling concept that also became culture.

If there was a pulling, and the word orbit cancels such an idea, then there has to be some sort of prevention taking place that disallows the pulling to commit the direction of travel. I know it is said that the orbiting object falls as fast as it circles and by falling while moving to the following side on position it never reaches the Sun, and yes, it makes sense, but there has to be some form of resistance replacing the planet in the next side position and preventing the falling or the pulling from taking place. If Newton is correct, then at least the circle must reduce, which it is not doing, the circle increases.

Using the formula $F = G \dfrac{M_1 M_2}{r^2}$ as Newton provided, disallows any other concept other than moving towards. The person Newton got his ideas from and the work he raped completely, that of **Johannes Kepler** who explained this very well, but **Johannes Kepler** makes no room for any pulling of any sort. In the work of **Johannes Kepler,** he said that the space being that which forms the orbiting route a^3 remains at a specific distance **k** while the orbit T^2 takes place…and in all my other books that addresses more information, I take Newton to task on his dismembering of **Johannes Kepler**'s formula by corrupting **Johannes Kepler**'s work and with such deliberate corruption, Newton's actions is nothing less that what amounts to fraud. By raping **Johannes Kepler**, Newton takes science on a goose chase that holds no truth. There is no pulling by mass of mass resulting in destroying the radius between any and every object in any way.

We have either one of two thoughts in the way we speak that has to be correct. Either it is planets that circle or as Newton would have us believe, it is planets that pull. If Newton is correct, then the normal way of speaking is incorrect. Then we must start saying planets are pulled to the Sun and not planets are orbiting the Sun. If you regard this idea as trivial, then you think of the entirety of physics as trivial. If the normal form of speech is correct and the planets are merely orbiting the Sun, then Newton is wrong. The planets can't orbit the Sun while at the same time we have Newton's accepted scientific presumptions being correct. The fraud part is in the accepting of the Universe expanding while still insisting that Newton is correct in his dogma of contraction with mass. This web page is an effort to show how Mainstream Physics brainwash students into accepting Newton's hypotheses of mass attracting by force while the entire Universe is expanding at the rate the Hubble Constant indicates ever since Big Bang took place.

Students, read the following message about my book I named ISAAC NEWTON: A CONSPIRACY TO DEFRAUD SCIECE and learn how you are brainwashed and how your mind is pre- conditioned into believing in Newton's myth of pure deception which Academics call physics. If you are a student in physics who don't believe that you are subjected to unlawful brainwashing, then read on and prove I am incorrect. All you have to do to accomplish proving I am incorrect is to prove that Isaac Newton is correct!

Let's start surveying civilized principles by evaluating what lawfulness means and what would constitute as morality. Let's determine what characteristic features makes the crook in the book?

If any person, notwithstanding what reason is given in justifying such depravity, tells a lie or conveys untruths to further whatever humble cause, it is seen as fraud. To convey information that is not substantiated as a verified fact then the mere conveying of such information becomes fraud.

When any person, notwithstanding what reasons given, repeats such a lie unabated while being well aware that the information passed on by such a person is incorrect, then the person commits deceit. When anyone is repeating the information that is passed on as being unblemished factual substantiated and verified truth while such a person knows very well that such information is void of proof or lacks proof, then committing such an act is a criminal enterprise. Academics in physics commit every one of the above indignities and yet see their actions as being lawful and even much praiseworthy and hold their role in society in the highest esteem imaginable. They fail to see the crime that they commit while tutoring physics. Whatever motivation they may claim to have which they offer to be serving them as forming their driving force, the fact that they perpetually perpetrate in unlawful behaviour, by spreading untruths, such actions on their part put those academics holding such highly regarded positions in the league of ordinary cheats, gangsters and common criminals. By wilfully and constantly falsifying facts to further whatever humble cause and produce illegal claims repeatedly, remains derogative behaviour and is unlawful by nature, notwithstanding what morality it should serve. A Preacher or Pastor lying on behalf of God is not lying on behalf of God and to think the Preacher or Pastor improves or underlines the Greatness of God by lying on behalf of God is very mistaken, because in reality such a Preacher is falsifying the truth for his or her personal benefit. Lying is wrong and doing so even in the name of God remains despicable. The same applies to academics in physics. There is no argument that can change this truth about falsifying the truth and when doing so there is no hiding behind any excuses of ennobling to benefit mankind that will change such truth into righteous conducting.

Newton said centuries ago that gravity is the force of attraction there is between objects that hold mass and it is the mass factor that brings about this attraction, which Newton claimed there is. The Universe does not contract and all the proof we require to disprove such a statement we find in the Hubble constant as a guarantee. Moreover, it is true that the Universe never contracted even for a brief instant and proving that is the Big Bang concept with all the proof that this concept brings in backing the principle of expansion in the Universe. Planets never moved closer, are not moving closer and will never move closer to each other and this is backed by all information collected this past century. The Moon is not coming closer but the distance between the Moon and the earth is widening. Studies about the Universe reveals every time that space in the cosmos increases constantly. Studies find all things are moving apart and away from one another. Any and all the proof about this is beyond what any doubt may present to counter this knowledge. Notwithstanding this irrefutable findings, science still regards Newton as the only person that ever lived whom no one ever could prove wrong…and this is upheld by Mainstream Physics in spite of the cosmos proving Newton wrong every instant in time. The basis of what science holds as its foundation we find to be the Newtonian principle of $F = G \dfrac{M_1 M_2}{r^2}$. The foundation used by science promotes this argument and backs up this argument well knowing that in the cosmos there is no evidence backing up this proposal Newton suggested. The Newton formula $F = G \dfrac{M_1 M_2}{r^2}$ used as basis for science sees gravity as being a force of attraction and the force of gravity is being in place between all objects in accordance with the mass factor that the objects have as presented by Newton in the formula $F = G \dfrac{M_1 M_2}{r^2}$.

What we find as we gauge all evidence found while studying the Universe, is that reality shows there is no attraction between objects in space going on anywhere in the Universe, that the entirety of such a concept is a myth and the outward moving of the Universe has been coming from and since the time of the Big Bang and maintaining this flow of material is substantiated in a concept named as the Hubble constant, which proves Newton's perceptions to be a myth. The Hubble constant proves that space everywhere is growing ever since time began and the growth never stopped ever since.

Knowing this irrefutable fact does not deter science from under scribing Newton as the sole basis that underwrites all the correctness of all of science known as physics. However, Hubble and the Big Bang and all other investigations contradict this attraction idea forming whatever Newtonian dogma holds. Therefore, any further believing that there is attraction going on as Newton claimed has to be viewed for what it is and that it is a fairy tail. The Big Bang Theory proves Newton's idea as not only being wrong but Newton's idea of attraction is a joke. If the Big Bang is expanding the Universe, then how can the Universe contract at the same time? Any contraction by nature would have the Universe collapse back into infinity the moment the Big Bang moved out of infinity. Ask your professor to show how an expanding Universe can also contract and your professor will tell you about Einstein's Critical Density theory. This theory I prove is the biggest fraud ever devised by any group of persons in the history of civilization! This is perpetrating fraud and conducting in upholding deceptions instituted by Newton that then formed the institution of lies they call physics.

The Universe does not contract in any way, means or form and even such a suggestion is incorrect! The Moon and Earth are not moving closer but are moving apart. The entire Universe is growing in space and nowhere is space depleting by any norm used. Academics are very aware of this misconception Newton had and still academics in physics are promoting the ideas of Newton as the only believable unwavering truth. Academics teaching these misconceptions are committing fraud, notwithstanding the portraying of their role in society being unblemished, spotless while they are covered in a lily white blanket making them being whiter than snow and having such a holier than thou attitude. Teaching Newton is participating in deception and promoting Newton is criminally deceiving the public and while anyone is doing so, such a person is committing an act with criminal intentions.

Then, in the face of all this evidence contradicting Sir Isaac Newton, they remain upholding the correctness of Sir Isaac Newton and keep on teaching students about the unwavering correctness of Sir Isaac Newton. They put down conditions of learning to this effect and are expecting students to repeat these untruths and unproven facts by forcing answers to that effect in examinations. Students are forced to believe that $F \ = \ G \dfrac{M_1 M_2}{r^2}$ holds all the correctness there are and must support this fact in every examination they take or be expelled from the course on the grounds of failing the examinations. Forcing the acceptance of this untruth about physics is equal to preposterous subjecting students to physiological torture and heinous mind conditioning, scandalous thought control and brainwashing for every aspect in the Universe contradicts $F \ = \ G \dfrac{M_1 M_2}{r^2}$. This applies to everyone serving as a tutor in physics notwithstanding whatever status the tutors or should I say the torturers might have in society or the morality they attach as a reason to commit such atrocities.

If you are a student, then you are conditioned by academics in controlling your thinking by enforcing pre-mind setting and in which they methodically force you into believing in Newton's formula being utter correct $F \ = \ G \dfrac{M_1 M_2}{r^2}$ and this is an on going process conducted for centuries in the past, while it is the truth that Newton is completely void of any tests that may secure any form of confirmation and in securing proof then also by that establishing proof. Read this book **ISAAC NEWTON: A CONSPIRACY TO DEFRAUD SCIECE** and then use the information I supply in the book to insist that Academics who are teaching physics, prove to students that Newton's statements of attraction are correct. Let those academics explain the method mass uses to initiate the attraction they claim is there. Let them, with precise detail show how attraction comes about when mass is applying, and that gravity is produced by mass and such producing of gravity then would establish attraction! I show precisely how gravity produces mass but I also prove that mass can never produce gravity. I show with explicit detail when, how and where gravity forms mass but mass can never form gravity. What I prove annihilates every Newtonian claim. Yet, since I contradict Newton (as the cosmos also contradicts Newton) my work is ignored by every person forming a part of or holds a position in Mainstream Physics.

Mainstream Physics never prove Newton's philosophy on gravity $F = G\dfrac{M_1M_2}{r^2}$ but those persons conducting teaching in the subject of physics force all physics students to learn that Newton's gravitational concept $F = G\dfrac{M_1M_2}{r^2}$ form the basis of all physics and to accept the facts as if it has been proven beyond all other facts. Students have to believe that Newton is correct or academics will see to it that they fail their examination and are then as a result removed from the institution of learning. If that is not mind control and thought processing, then what on Earth defines brainwashing better? The condition of being accepted in physics is to accept Newton without questioning or insisting on the proof that is never supplied.

Let those academics now prove precisely how mass brings about gravity and then afterwards test you on how Newton is proven correct and not on you repeating facts about what they say is true about what Newton said, which they say is true. The manner they present Newton is completely hearsay and that method may not be used in any court of law. Let your professors now prove how it is that Newton's teachings are correct while the entire cosmos is showing the very opposite and then examine you on the process they use to prove Newton's concepts. At present they say Newton is correct and then they test you on your ability in forcing you to repeat that Newton is correct without ever proving to you that Newton is correct. Let those physics professors now prove Newton and then test you on the manner they use to prove Newton to be correct. The truth beyond all other truth is that Newton's gravity has never been proven (because try as you may it is not possible to prove Newton's formula forming gravity mathematically) and because academics know that, academics require the blind acceptance of Newton by students. This unconditional acceptance in Newton's correctness relies only on the pre-conditioning of students' mindset and academics depend only on the student trusting the academic "say so" about the institutionalised correctness of Newton.

That Newton is correct nevertheless and notwithstanding that there is no founding proof about this matter, is what students should be accepting blindly, or be chased from campus. Pre-conditioning students into blind acceptance depends on the academics' insistence that students approve Newton's concepts without pre judgment or students insisting on scrutiny of any sorts. In examination students have to outright and blindly follow academics' say so only because academics say so. Academics depend on students never questioning their say so or demand proof about what academics teach. Those academics in teaching positions insist that all students accept Newton's accuracy. One such a proof is to show just how much is the Moon coming closer to the Earth per year!

This is methodical mind control as much as it is the brainwashing, while I show that they enforce the dedicated and the devoted accepting thereof. If you are one of those believing that Newton was proven, then what you believe to be true is a lie because Newton can't be proven and that is the truth! The time has come to face your teachers and force them to stop the ongoing old culture of bullying students and conditioning their thoughts by enforcing on them dogmas which is mind control! In order to get students to accept Newton's hypothesis, academics resort to brainwashing pupils and students by never giving any proof of how the force of attraction comes about. While I do just that, those same academics that are unable to prove gravity by attraction ignore my work to benefit their thesis wherein they hail Newton's greatness. They teach you that the Universe contracts and to state their case they force students to learn that gravity is proved by Newton introducing the following formula $F = G\dfrac{M_1M_2}{r^2}$ They say that M_1 is the mass of the Earth and M_2 is the mass of the individual in questions mass and the multiplying of these factors with the gravitational constant produces the force of gravity when this gets divided by the square of the radius.

Please let you lecturer put in all the values of the formula and prove Newton is correct. If he can't and I know for sure he never can fill in the symbols and calculate the force of gravity, then read the rest of the web page that follows to see how far academics in physics go to brainwash students into believing in Newton's fraud. This is a fair test to see if Newton's contraction theory underwritten by

Newton's attraction formula $F = G \dfrac{M_1 M_2}{r^2}$ is valid, then force your professor to use this formula as it reads and show WHEN the Moon and the Earth is going to collide. If he fails to do it by using Newton's formula as $F = G \dfrac{M_1 M_2}{r^2}$ then you will know who is conning you, him or I and who is truthful in informing you correctly, again I ask you to judge whether it is they or I. I charge all academics to prove what I say is being wrong in any way or even that I exaggerate in the least. I challenge Newtonian academics to prove that mass does indeed form any force of any sorts and in particular gravity! To those professors claiming Newtonian ideas are substantiated by proof, I say that notwithstanding your personal academic qualifications and while at the same time disregarding your status and previous achievements as well as ignoring your many admirable abilities you may have and however superior they might be, I shall teach you the correctness about gravity when you read my more informative work, such as An Open Letter On Gravity Part 1 + 2 Volume 1 +2. I say it is time students learn the truth about physics, notwithstanding the status value that academics will loose in the process. Students do read ISAAC NEWTON: A CONSPIRACY TO DEFRAUD SCIECE and challenge those academics that depend on their ability to brainwash you into submission. There is no sensibility and a constructive debate when the science involving the construction of the Universe is debated, but instead there is a constant conflicting of facts bringing continuous loose ends. Never is there direct correlation in science. We live in a cosmos that should attract but expands. We have an Earth that attracts the moon but never does the Moon come one centimetre closer. We have a Sun attracting planets that orbit without ever closing in on the Sun. In all attraction that Newton's gravity should have, we find planets constantly drifting apart.

However one wishes to look at it, the result is the same…there is something seriously the matter with science. If the Critical Density Theory that is supposedly holding the dark matter mystery as the answer, then why would the Dark Matter not generate gravity at present and bring contraction. If the Dark Matter is with mass forming gravity through the intervention of mass and holds space currently in the Universe, then what stops the **Dark Matter** from forming gravity by the way of attracting? While there is mass, whether the mass is with light or in darkness, the matter has to generate gravity, if there is mass and if it is mass that generates gravity. Being visible or unseen has no influence on being with mass or having the visibility preventing attracting gravity. With the mass being there, the mass has to induce attracting gravity. The Newton formula used as $F = G \dfrac{M_1 M_2}{r^2}$ does not allow for the radius to have the mass influences become stronger as the radius grows but mathematically shows quite the opposite will apply. With the radius growing the influence of mass will tarnish. Therefore, as the Universe expands, the influence of mass will reduce by the increasing of the radius.

In the presence of unseen dark matter being present, then why is the contracting Universe not applying attracting gravity at present, because we are in an expanding Universe, notwithstanding Newton's misconceptions! If mass do bring gravity and gravity is contraction or attraction, and then be the material in question dark or luminous, the material within the Universe has to produce enough gravity to bring contraction or else there is insufficient material to bring attraction, or gravity does not bring attraction as Newton supposed! There is attraction with the measured value of mass in quantifiable numbers or there is no attraction going on. However, if that is true that there is no attraction and only expansion in the Universe, then everything that science claims about fundamental facts in the Universe about the Universe is incorrect.

I do have the answer, but it is not what Newton said the answer is. By using Kepler and not the part about Kepler that Newton raped, I can show what gravity is and how the Universe unfold when we apply the sensibility that Gravity brings as Johannes Kepler introduced gravity. I wish to share the answer with the world, but everyone with any personal interest or any formal say in the matter, albeit a thesis they wish to protect, money they have vested in books and theories or an ego that will not permit them to acknowledge they are wrong, all the above mentioned has an interest in stopping my books being published. Even publishing houses has much to lose with loss of income from their other books they market when my work disproves everything that was written and in that I can find no publisher that shows any interest in having my books published. There is much incorrect about the manner in which Newtonian wisdom see the Universe. There is much incorrect that Newtonians wish to cover-up.

Answer this: If the Universe is expanding while Newton said the Universe is contracting, then why is the Universe expanding while the Newtonian gravitational wisdom says the Universe in gravity is attracting. When you wish to hide behind the hoax that is named as the Critical Density Theory, then answer the question as to why is the dark matter (if there is dark matter) not generating gravity when the mass must be present. If there is matter there is mass and if there is mass there has to be attracting gravity and with such mass present, why is the Universe expanding? In **Newton's Fraud** I show to what length science went to cheat figures on behalf of Newtonian views and just how far they are prepared to falsify the truth to cover-up the deception that Newton started. I show how the biggest fraud ever perpetrated is committed in the name of Newtonian science by presenting the corrupted arguments they have as so called proof which forms part of the Critical Density Theory. Read how students have been brainwashed for centuries to accept Newton and if you disagree or do not believe me, then use the Newtonian gravitational formula to show how the Moon is closing in on the Earth and at what rate are the Earth going towards the Sun since the Earth is attracted by the Sun.

What you are about to read when purchasing any of my work, the information was never printed. What you are about to read in my work diverts from "Xepted Science" (which is another name I gave to mark Newtonian science in order to distinguish between reality and what fits scientifically into the story of the Emperor's magic clothes), as much North diverts from South. The science you are about to encounter through investigating my work does to Newton's views what Nicolaus Copernicus did to Ptolemy's Universe model, as it pushes the old concepts out of the realm of reality. The work you about to discover, is not part of any encyclopaedia or accepted textbook. The aim of what you are about to read intends to explain everything Newtonian "Xepted Science" thus far was unable to explain. That which you are about to experience, "Xepted Science" science are unable to explain: The Titius Bode law; The Roche Limit; The Lagrangian Point System and; The Coanda effect is so far away from being explained by using Newtonian genius. Newtonian genius has condemned these existing phenomena found in the Universe as not existing or being fictional, that is the only way they can explain what forms the cosmos.

The phenomena mentioned is not necessarily explained in this book, but if it is not explained, then this book is doing ground breaking work to have the four phenomena, The Titius Bode law; The Roche Limit; The Lagrangian Point System and; The Coanda effect explained in the other book. The work is only explained in books such as the four volumes forming **_An Open Letter on Gravity._** As the phenomena don't support Newton's vision on cosmology, (although there is no other support or backing of mass by gravity anywhere in the Universe to confirm Newton's cosmic visions) the phenomena has no support amongst Mainstream science although they did apply it with many a positive result in locating the missing planets at the time of their discovery. But because explaining these new concepts that I introduce, involve facts that Newton never understood, this information is so new that you are most probably one of the very first to encounter the work. Do no look for any bibliography on this work. Do not look for other authors supporting these views. I can't give you any cross reference to read and have what I explain better explained by another person. If you wish to compare what you are about to discover with information previously announced as new work, then it will be best to look at my work on the day science takes on a new personality and casts away its Newtonian chrysalis jacket and a new butterfly called reality cosmology emerges. When looking for any sort of reference on my work then there is nothing to refer to at present in science.

My work introduces into cosmology what Newtonian science at this point still denies being present in the Universe as forming a presence as being a factor of any sorts whereas I not only mathematically explain the phenomena but also mathematically put it into context with gravity forming. Newtonians can't even recommend any person a book to read about what forms the phenomena such as The Titius Bode law; The Titius Bode law; The Lagrangian Point System and ; The Coanda effect.

The Titius Bode law; Newtonian science goes as far as condemning these phenomena as coincidental because they have no thought about what it presents, although they used it in the past to locate stars.

By using the above, the four cosmic pillars, it enabled me to present the proof, where I now can explain what conditions bring on the sound barrier.

Two persons of whom **Johann Elert Bode** was one of two persona that could legally lay claim of being responsible for discovering **the Titius Bode law,** which I present as one, of **the four pillars** on which the Universe rests. He was born in Hamburg Germany, in 1717, and published his first astronomical paper in 1766 on the eclipse of the Sun that took place in August of that year. This paper was followed, two years later, by an elementary text on astronomy, which led to his appointment to a board for the improving the yearly tables of positions and planets.

He founded the *Astronomichs Jahrbuch,* the widely known German publication, and edited it for 51 years until his retirement in 1825. He was made director of the Berlin Observatory in 1786 and dies in 1826. In 1772, Bode brought to the general attention of the world a law Newtonians dispute as a law but is a law notwithstanding. The law accounts for a definite an undeniable accounting for the relational orbiting of planets orbiting the Sun. The law takes its name from the two persons that made the discovery where the other person is J. D. Titius, of Wittenberg, which was revealed several years later.

At the time, the three telescopic planets, Uranus, Neptune and Pluto, had not been discovered, nor were the asteroids known. Indeed, it was the Titius Bode's law, which was employed to point the possible allocated positions for the discovery of the asteroid. Bode as well as Titius took a set of numbers, beginning with zero (which should be infinite) and increasing geometrically in steps of 3: 0, 3, 6, 12, 48, 96, 192, 384, 768.

To each of these numbers they added 4 and then divided the results by ten. This gave them a series, which matched rather closely the distances from the Sun of the planets which has always been known to man, when the distance were given as astronomical units. An astronomical unit is the main distance between the Earth and the Sun. Because the Earth's orbit about the Sun is not a perfect circle, but an ellipse, the Earth is considerable nearer to the Sun at one time during that year than at any other time, and it also reaches a point where it is the furthest from the Sun. All of this I dispute most strongly but that is for another time in another book. The Sun – Earth distance is 149 600 000 km and that is considered as an astronomical unit.

Bode's law also known as the Titius Bode law

A number series that roughly coincides with the average distance of the planets from the Sun outward to Uranus. It is named for the German astronomer Johann Bode, who pointed out the relationship in the 1772 edition of his book Introduction to the Study of the Starry Sky.

According to Bode, if the distance from the Sun to Saturn (then the outermost planet known) is taken as 100, then Mercury is separated from the Sun by 4 such parts.

The distance to Venus is 4 + 3 = 7,

to Earth 4 + 6 = 10,

and Mars 4 + 12 = 16.

Then at 4 + 24 = 28 comes a gap.

Jupiter is 4 + 48 = 52 parts away,

and Saturn 4 + 06 = 100.

Planet	Mercury	Venus	Earth	Mars	Ceres	Jupiter	Saturn	Uranus
Bode's Law distance	4	7	10	16	28	52	100	196
Actual distance	3.9	7.2	10	15.2	28	52	95	192

When the planet Uranus was discovered in 1784, it seemed to fit this scheme, though not with perfect accuracy. Bode's law led to a search for the missing planet between Mars and Jupiter, resulting in the discovery of the asteroids. The law was also used as a starting point by astronomers in search of planets beyond Uranus. However, it clearly breaks down towards the edge of the solar system. According to modern views, the planets would have come naturally to take up station in orbits where they least perturbed each other; such an arrangement would spontaneously produce some kind of Bode's law progression of distances in any planetary system.

Bode's "law" is thus not really a law at all. What is more, it was first put forward in 1772 by the German mathematician Johann Daniel Titius (1729-1796), and is now often referred to as the Titius-Bode law.

When adding the four to the three one gets to the number of seven and where space reverts to three points we find time referring to four. That makes the total value of motion seven in relation to the time movement, which is ten where the total relation is three (3, which is the space aspect) and four (4, which is the time or rotation aspect) and that becomes the seven indicating a sphere. Later I will give more detail about how this develops into the mathematical value of gravity. In the table above a close agreement with the actual distance is shown where it helped in locating the unknown planets that were discovered later on. According to legend the discoverer of Ceres, Giuseppi Piazzi (1756-1826), was combing the sky in search of the real distance between Mars and Jupiter while keeping the Titius Bode law well in mind. Uranus fall beautifully in place, and it was not found until 1781. Neptune is somewhat out of line but Pluto again falls much better into the slot reserved for Neptune. This I do explain mathematically in **Seven Days of Creation** and also in that book I explain why the four inner planets are terrestrial planets and why Jupiter is the giant it is.

The Roche Limit; the photos indicating this process happening they explain as "stars blowing bubbles"

The Roche limit is:

The region surrounding each star in a binary system, within which any material is gravitationally bound to that particular star. The boundary of the Roche lobes is an equipotential surface, and the lobes touch at the inner Lagrangian point, L_1, through which mass transfer may occur if one of the components expands to fill its lobe. It names after the French mathematician Edouard Albert Roche (1820-83).

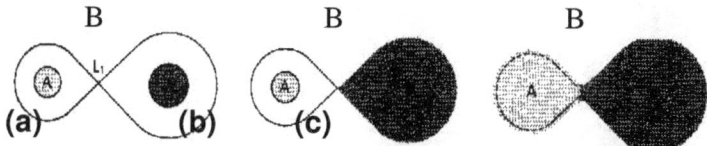

THE ROCHE LOBE: In a binary system, the Roche lobes of components A and B meet at the L_1 Lagrangian point. (a) In a detached system, neither star fills its Roche lobe. (b) In a semidetached system, one massive component, B, fills its Roche lobe. (c) In a contact binary, both components overfill their Roche lobes and share a common envelope. As with the graph I can see the two sides forming a connection therefore relevancy has to apply, all contradicting Newtonian claims of no connection but through mass attractions. The mass does not attract but one interferes with the other total influencing the space surroundings.

Considering Official Science policy the collision must be devastating and total destructive to one or both. Where r that is the radius between the two colliding structures disappears from the equation since the collision is already in progress the structure would be unable to maintain any viable distance.

$$F= \quad (M \times m) / r^2$$

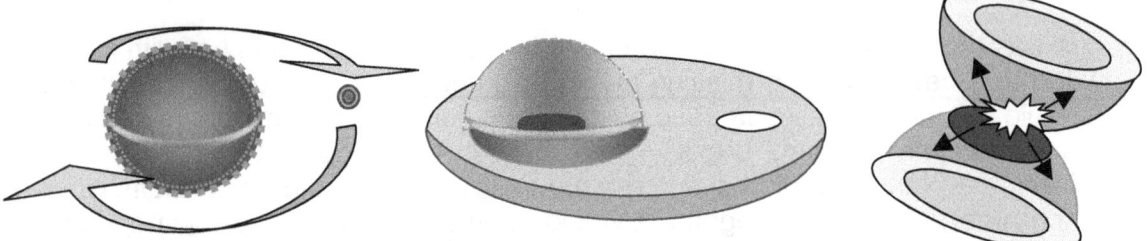

Even more astonishing is facts about the Binary star system that is seldom to never mentioned. The fact that Binary stars form and never collide disproves Newton's attraction theory completely!

The second pillar that I introduce is named after **Edouard Roche** who was a French astronomer who did considerable research into the nature of the Earth and the other planets, particularly as regards to their construction. He proved in 1850, that the liquid satellites of any planet, if at a distance from the planet is greater than a specific limit, would be distorted by the tide raising forces resulting from the planets gravitational attraction. This I dispute. It has all to do with singularity being penetrated by a solid that turns into a liquid. If, however, the satellite were nearer to the planet than its critical distance, the tide-raising forces of the primary would overcome the mutual gravitation of the satellite's parts and would liquefy the satellite. If the planet and the satellite were the same density, the critical density, known as the Roche limit is 2.47 times the radius of the planet. If the satellite, or the material of the satellite might be composed is a solid rather than a liquid, the forces of the planet will prevent the coalescence of this material into a satellite and force it to remain in fragments which would assume orbits about the planet according to Kepler's laws relating to the mass of each fragment.

THE ROCHE LOBE: In a binary system, the Roche lobes of components A and B meet at the L_1 Lagrangian point. (a) In a detached system, neither star fills its Roche lobe. (b) In a semidetached system, one massive component, B, fills its Roche lobe. (c) In a contact binary, both components overfill their Roche lobes and share a common envelope.

         The Roche limit is:
The region surrounding each star in a binary system, within which any material is gravitationally bound to that particular star. The boundary of the Roche lobes is an equipotential surface, and the lobes touch at the inner Lagrangian point, L_1, through which mass transfer may occur if one of the components expands to fill its lobe. It names after the French mathematician Edouard Albert Roche (1820-83).

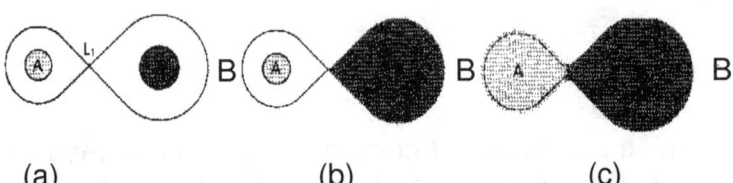

    (a)   (b)   (c)

By using the four pillars, it becomes possible to mathematically prove gravity is generated the same way as electricity is generated and there are no differences between the two. The motion of the Earth in relation to outer space generates gravity. When applying motion apart from the motion that the Earth provides, the individual structure that is producing individual motion within the sphere of motion that the Earth's gravity provides, is in access of the gravity the Earth provides. Such extra motion becomes the independent and individual motion that puts the relevance gravity has beyond the conserving means that the Earth's gravity has, where the space that is serving the independent object is independently in motion and is apart from the motion that the gravity of the Earth provides. By applying individual motion that increases the motion the Earth provides the independent object, then the individual object is becoming further independent. This continues to such an extent where the motion creates almost the ultimate independence that is breaking the restraint gravity has on all objects with independence within the sphere of the Earth's gravity. Breaking the sound barrier is the motion in space duplicating space by crossing over gravity borders.

The Lagrangian Point System

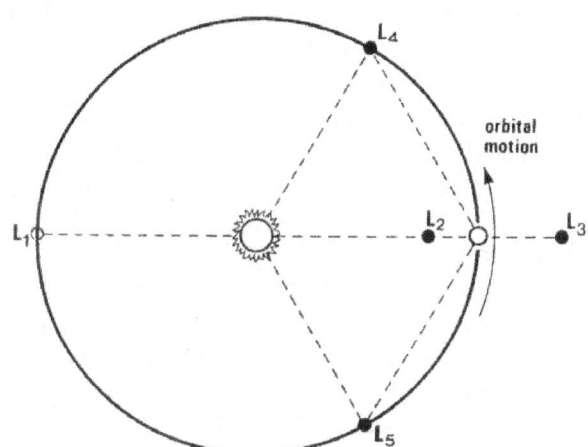

There are three other cosmic pillars on which gravity is founded and inspecting any or all of the others there is not a single bit of evidence of mass playing any part in the overall concept. They are the Roche limit, Lagrangian points, and the Coanda effect.

The Lagrangian system is in its full compliment the Coanda effect that derives from the Titius Bode law which presents the best example of the Coanda Effect establishing gravity by lining up liquids in relation to solids rotating.

Lagrangian points: Five points in space at which a very small body can remain in a stable orbit with two very massive bodies. The points were first recognized by Joseph Louis La Grange and are rare cases in which the relative motions of three bodies can be computed exactly.

The Lagrangian Point System   They can't deny this taking place because the evidence of this happening is overwhelming but also they haven't got a foggy clue of why this is happening. I not only explain why but mathematically use this as being part of the process forming gravity.

…and; then last but also culminating the previous three in forming **The Coanda effect.**

I am casting aside four hundred years of nonsense and ideology that has no fundamental support in the cosmos. I dispute Newton's claim of $F = G \dfrac{M_1 M_2}{r^2}$. How much further can I go back to dispute current physics ideology and Newtonian religiosity? I dispute the very essence of mass forming a contraction which is called gravity and the best available explanation for the presence of gravity is on the grounds that it works by a magical occurrence where one body can grip onto another body and pull…without having physical contact in any way at all. In contrast to this bullshit I use the Titius Bode law whereby which the entire solar system was formed and in relation to the Law of Pythagoras I show that gravity takes place by the movement the Earth (or all other cosmic bodies) exerts. But gravity by mass can't begin to explain the Titius Bode law and therefore with their Newtonian brilliant thinking they deny the reality of the Titius Bode law being present and being a law. Instead they blame the cosmos of being incorrect by expanding to justify their claim that Newton could never be wrong.

Look at my work as if the story of the Bible is happening at the point where the Israelites were to be rescued and at the very spot in time when the work as the work of Moses as he came down to Egypt to rescue the Israelites and that was before there was a written Bible of any sorts. I do not compare myself in person to Moses, or compare my work to that of Moses, but am trying to establish a reference about my work to science as one would start to read the Bible.

The problem that science at present experiences is the absolute blindness is self-righteousness and mania in personal competence physic academics think they have and this inspires a feeling of megalomania that Newtonian academics experience concerning their work. To them Newtonian science is all there is because to their simple minds Newtonian science constitutes what there is in the Universe they have. They have painted their view in a flat corner. If anything doesn't support Newton then everything including even the cosmos has to be incorrect and anything not full-heartedly supporting Newton's infallibility should immediately be corrected. To them Newton is correct and anything that strays from a thought that Newton is correct should be corrected because only Newton could be correct.

They don't even try to explain what Newtonian science can't explain and they would even put the cosmos at fault if the cosmos tries to expand while Newton explicitly said the cosmos is contracting. In that case they will launch an investigation, not to see where Newton faltered, but to see what the Universe will do in order to correct its unacceptable diversion from what Newton specified as correct. The cosmos must correct this expanding and subdue its rebellion to convert the expanding to contraction and if need be, then fictitious dark matter has to be located in order to correct the cosmos' silly ways. If Newton can't explain **Titius Bode law; The Titius Bode law; The Lagrangian Point System** and; **The Coanda effect** then at best it doesn't exist and then it remains the best probability to be left as if it never was.

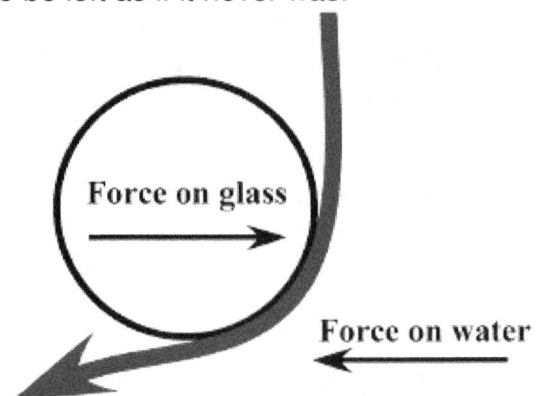

Force on glass

Force on water

Newtonians declare that there are four official forces that the Newtonian wisdom invented as a compromise for the Newtonian failing to understand science. However, when required and if and when need be, Forces could be called on at will to help hide more Newtonian ignorance and protect the detecting of much Newtonian incompetence about understanding the science of cosmic physics. See how quick more forces come to their rescue when called on to be used in explaining the **Coanda effect.**

     I don't only claim there is a reaction for every action as Newtonians do, but I also show why this is true. I show and why Newton's three laws on motion apply and not only indicate to these laws being present! I use cosmology and mathematics to support the three laws on motion to a far better degree than Newton's feeble mentioning of the laws…but when saying this don't expect to find all the answers of four hundred years of mistakes being corrected in a few pages of easy reading relaxing entertainment. My work doesn't relax. My work doesn't entertain. My work doesn't compromise. My work goes much further than informing; it explains by educating but in doing that the level of understanding that is required to grasp goes very deep in some cases. When you read my work you will be in a position that the explorer Vasco Da Gamma found himself when he explored the African route to India and discovered the Cape of Storms. Every page will be very new and exceptionally demanding in concentration since as you read the information, you will most probably be the first person you know that would embark on this scientific discovery. Enjoy the journey.

This was

www sirnewtonsfruad. Com

Are you at this point still very convinced about how valid and reliable would the **Newton** gravitational formula

$$F = G \frac{M_1 M_2}{r^2}$$

is and also remember that on this formula rests all the fundamentals forming every principle attached to physics!

If this formula is as valid and as reliable as students are brainwashed to accept that it is, then are they able to measure at what a rate is the Moon coming closer to the Earth? Surely those person's with smart enough abilities being able to calculate physics and dabble in rocket science should be able to calculate an approximate date when the ultimate Earth /Moon collision would take place by using the all so accurate Newton gravitational formula

$$F = G \frac{M_1 M_2}{r^2}$$

and then permitting the absolute accuracy of such reliable calculation we can school the thought of when the end of the Earth would take place?

If you are one of those members of society who never thought you would hear the name of an accomplished person such as Sir Isaac Newton being associated with fraud corruption and brainwashing, then these books are specially written to inform you about the truth there is lacking in the correctness about science. Planets do not pull and therefore there is no attraction between planets and or the Sun. This is why…

Are you under the impression that everything Newton ever surmised about gravity is proven beyond doubt?

Are you under the impression that everything they teach about anything Newton ever surmised about gravity is proven is unblemished and is also completely beyond doubt? If you are, then also you are in for a nasty wake-up shock.

The definition of gravity is that gravity is a force of attraction between bodies with mass…and if gravity is a force of attraction, then when will the Sun collide with the Earth? If gravity is a force of attraction, then when will the Moon come crashing onto the Earth?

How valid and reliable would you say is the Newton gravitational formula

$$F = G \frac{M_1 M_2}{r^2}$$

and also remember that on this formula rests all the fundamentals forming every principle attached to physics!

Does any of the answers to these questions interest you…then please read on for there is much more. If you are tickled by the intrigue that one would find in bringing the Newton gravitational formula

$$F = G \frac{M_1 M_2}{r^2}$$

to any conclusion, then go on reading this web page. If you are a student in physics at any level or doing any studies founded on the accurateness of physics, then you better read this web page to find conclusive answers to defining questions about physics and the accuracy of what you study when you study physics!

I wrote many letters to the addresses of many academics while never receiving any answers about the facts I showed that was in my view discrepancies in the principles of the work of Sir Isaac Newton. Then the final letter came that I wrote to the address of any academic, which the book entitled An Open letter Announcing Gravity's Recipe holds this final letter to such academics and forms the first 100 pages of he above mentioned book.

That was the last letter I intentionally ever will write to academics. In usual letters I normally address to academics I use the profile that might be on par with work that will attract the attention of such esteemed intellectual academics. In that final letter that I sent to the addresses of such academics, I warned those I contacted that I aimed to uncover their Master's fraud, a person called Sir Isaac Newton and that I am going to accuse him as well as them of fraud.

They never took me seriously.

I warned them I was going to show just how, why and in what way do they brainwash students in accepting the principles that support Sir Isaac Newton and how they place mind control on students to further the students' acceptance of Isaac Newton's physics principles without asking questions.

They didn't take me seriously and now I am taking them to task. I am fulfilling a promise I made some time back by showing how and why I accuse the physics establishment of committing blatant and total corrupted fraud.

After I completed An Open Letter Announcing Gravity's Recipe, I realised the trend was somewhat on a higher level than what the average student may appreciate and another more simple book was needed to inform students. From this realisation I went about to form a framework wherein I highlighted the incorrectness there are in the work of the person known as Sir Isaac Newton. What I wrote was meant to be a framework and no more. But as the book developed, the book became progressively more complicated, bigger and more informing. Taking on Sir Isaac Newton and his order of high priests always leads to more explaining about increasingly more complicated and more complex issues, especially where one throws out the magic Newton put into physics to explain forces and other mythical ideas that belong to the Dark Ages, such as the concept that gravity is being a force of attraction. I wrote a framework with seventeen divisions, which form the four books entitled an Open Letter On Gravity. However, after completion I found the book having about one thousand seven hundred plus pages (in manuscript form) and the book in that form was far too heavy to handle and cumbersome.

Also I became aware that the book an Open Letter On Gravity again developed in something students might not appreciate that much because again it turned out to be very informing but once again rather complicated, or too complicated for students to enjoy as easy reading material.

With a book being several kilograms, I was forced to split the book into four equal part naming the one an Open Letter On Gravity Part 1 Vol. 1 + 2 and the other an Open Letter On Gravity Part 2 Vol. 1 + 2. Since the content of these volumes forming an Open Letter On Gravity Part 1 + 2 Vol. 1 + 2 also turned out as seeming complicated or complex, I decided to form a book from what initially was the framework a book that showed Newtonian Fraud, which I subsequently named as such. Then I decided to make the book into several titles bringing a level of choice to the one book where each title holds the same information as the other books, but the books become more informing and more complex with the adding of more chapters. Every book in the following that I offer is the same accept for the adding of chapters to increase the information and intensity of explaining, as well as the level or degree of technical understanding the book requires. The first one of the books that I offer holds 5 chapters as a choice and the sixth book holds 14 chapters. I have given this choice so that the potential reader has the choice about how much information the person wishes to purchase. Please also take note of the fact that the degree of information and the complexity in explaining increases with the adding of chapters. The level of understanding the arguments require that I bring forward as well as the level of complexity portrayed by the arguments increase with the number of chapters increasing in every book because the intensity in the explaining level rises with the adding of chapters. Nevertheless, even with the book forming fourteen chapters, the intensity of the fourteen chapters does not remotely compare to an Open Letter On Gravity as it remain only a framework for an Open Letter On Gravity.

Touching other matters that I also address and confront by explaining:

How can a Universe contract as Newton said it does while that same Universe is expanding? In accordance with the Big Bang theory the Universe expands while according to Newton the Universe is contracting? If the Big Bang is correct then Newton is wrong and if Newton is correct then Hubble expanding and the Big Bang is wrong.

How can there be a Universe that holds everything there ever can be and has whatever could be expand and therefore become more while the Universe already holds whatever there will ever be? To expand something has to increase, in order to expand that which already is. What there is must then become more of what there is in order to hold more of what already is available. The Universe already holds everything there will ever be, so what can become more in order to expand? If it is understood that it is space that increases and space is filled with nothing, then how can nothing become more when nothing is the defined by what is the absolute absence of everything. To have "nothing: as a filling of tangible substance in the Universe becoming more in order to have the Universe expanding, then the Universe must reduce what it already has to increase having nothing whereby nothing then becomes more because more nothing applying means the removing parts of what has tangible substance of some of what already is. How is it possible to have "nothing" filling the Universe and still have distances between objects? It is said by those very smart and distinguished brilliant academics in astrophysics that there are 149 X 10^9 km of *nothing* (please note the word *nothing* that is specifically used) to hold any distance. The word *nothing* states a detail of what are total absence and the lack of anything present. How can that which can't be present because it is absent such as the term *nothing* does specify, how can that then fill space in terms of distance measured in kilometres or astronomical units. If there is 149 X 10^9 km of *nothing* filling the space all the way from the Sun to the Earth, then how long is one specific point holding *nothing* in terms of a measurable unit?

How is it possible is that mass can be responsible for objects falling by creating gravity by which objects supposedly fall and then have all things fall equal as Galileo proved. All things do fall equal...

If you think this is nonsense, then you are reading official Mainstream astrophysics. You will read about things they find "on the edge of the Universe" while everyone knows if the Universe does have an edge the Universe must end there and any end must have a new start. How could the universe that can never end have an edge where having an edge means it must end and having such an end is having a start? Whatever ends must start to bring an end and the Universe can never end.

However, please note that this following information mentioned is supplied as mostly being a part of the four books entitled as

An Open Letter On Gravity Part 1 and 2 Volume 1 and 2

How and where and at what point can we see did the Universe begin... and if you think answering this is impossible you haven't followed the trail of thought that Kepler left. You were only misguided by Newton's misinformation. Why would you be most likely correct when saying the Universe is a sphere... as it always is depicted in pictures?

Have you thought about the following? Because these I can answer...

...where is the cosmos coming from...?

...where is the cosmos going to...?

...and most of all...

Why is the cosmos travelling through time ...?

...what brings about the direction of expanding?

... (And it surely is not because the nothing filling the Universe is becoming more...)

My studying Kepler helped me finding answers to all the questions, which was deemed impossible to answer. The following is only a few of the many questions that I do answer.

Where is the centre of the Universe?
How did time and space begin?

That I can answer...and I also can answer...

...why is the Universe still growing since the Big Bang...?
...why did the Universe start so very small...?
...why did the Universe fit into a neutron at one time...?
...how did everything expand from fitting into a neutron...?
...why does space grow from small to large?
...where is it going while it is growing ...?
...why was the Universe any specific size...?
...what was everything before the Big bang...?

I grew tired of apologising for my (as they see it) having the audacity of being correct on matters of Newton's incompatible religiosity which I bring to their attention. When being in contact I am expected to show the utmost humble attitude acknowledging their supreme posture with me being in their surreal presence. I have to feel honoured to be in their presence when I mention to them their mistake about Newton being mistaken about a Universe that is contracting according to Newton while it never ever contracted in the least.

I am quite fed up with the attitude of those academics looking down their noses at me or worse still are those ignoring me whenever I show that Newton's facts just don't add to a conclusive believable answer.

I am at my limit with being ignored because those academics can ignore me by using their all powerful status and with them never having to prove Newton and therefore disclaiming even my presence when I try to disprove Newton.

I have reached my peak with stomaching the corruption they hide behind a lily white cover of dishonesty while they sit in their mighty towers and live in a bubble where not even God can touch them less having me point a finger at their despicable ignorance about their mistaken Master they portray as a God.

They can say the Universe is made up of nothing and go unchallenged for making a most senseless statement and when I bring this to their attention in a book, I am the person they condemn as being incoherent with my arguments about their nothing they have in place.

I say this again: any person telling a lie is committing fraud be it in the name of God or of science; such a person is a despicable liar. When they tell a lie to distort the truth and find financial compensation while falsifying facts, even if it is in conducting science then they are behaving criminally. That is distortion and is equal to the behaviour of the Mafia.

From my view and from my perspective, I honestly can't see any difference between the Mafia's racketeering and corruption and what academics commit in the name of being honourable scientists.

Those in charge of Mainstream physics feed students lies in order to be compensated for their misrepresentation of the truth. They are being paid enormous salaries from student fees to ensure that students believe in the impossible and accept what can never be proven and force students with methodical examinations to repeat the unproven or be expelled from the institutions and branding those expelled students as failures. That is a rip off whether it is justified as science in the process of learning or if it is plain legal criminality; it remains the same because they fly the same banner.

Those academics in key positions of academic credibility keep certain facts and evidence away from students and give other facts that were never proven before, prominence as well as credence while applying their trade in brainwashing to give their Newtonian views undeserved credibility and from these proceedings they earn substantial incomes. That is the same as racketeering. When you deceive by conveying untruths and cheat to mislead, then your behaviour is criminal.

If you purchase any of the following books you will come in contact with the truth for the first time in centuries. My work is about uncovering the truth and blaming the shameful conduct of those persons no one expects to be criminals.

When you purchase my books, I don't sell ink on paper. I do not sell material with questionable information, holding facts that were repeated so many times that it is accepted as the truth because it became a culture to believe Newton.

The information my work caries, which you will read in the event of purchasing my books, you have never seen, it was never yet mentioned or the facts I divulge has never been printed by any person, ever before.

I put untruths about the work Newton claimed as correct in question and that might have been published before but never published as questionable evidence. The rest I bring is new.

The following books that I offer for sale in this web page are unique in every sense. Only the information I question has been published before. The books take on Mainstream Science, uncovering facts that were never touched by any person for more than three hundred and fifty years. I act on behalf of the students in protecting those students who at this time are studying, or those who studied physics in the past. I am giving students information that is hidden by Mainstream Science.

On the book entitled ███████████████ I am making little profit. I give the book at a price that is going covering basic cost to furnish students with facts with which they can challenge academic torturers. I sell this at cost to show I am truthful in the hope of selling my other books.

The book entitled Newton's Fraud is showing more profit. The profit is going towards my attempt to get the books published privately which is done at a considerable cost. As the Purchaser will see, there are a significant number of sketches in the books and private publishers charge money for every sketch printed in the book. It is said that a picture is worth a thousand words and with the explaining I offer this saying is certainly true.

The book entitled Sir Isaac Newton: a Conspiracy to Defraud Science is showing much profit and that profit is going towards covering the expenses of what the publishing costs will be that this book holds. The information in these books mentioned above has never been published but moreover has the information appearing in the book Sir Isaac Newton: a Conspiracy to Defraud Science seen the light of day. In the first two books I deal with the problems academics in physics are hiding whereas in the third book I bring the solution to the problems. This information has never been in print or given in printed form to any member of the public and that information I do not intend to divulge free of charge. Should you wish to know what the problems are in the attraction by mass theory of Newton, and you wish to learn the truth about the working of physics plus you wish to find the answers concerning the truth about physics and in particular gravity, then you will have to pay for learning my decades of research.

All proceeding will eventually go towards having these books privately published and have the information freely available to members of the wider public. I have run into a brick wall called Mainstream physics and there are (I suppose) hundreds of reasons for preventing me from having these books published of which no less is the finically motivation to restrain the publishing of my work since this work will condemn many other books with profitable titles to the incinerator. If my books do get published, thousand of books that are already selling on the commercial market will have the same informative value as the fairy tale story of Cinderella. However, publishing costs are astronomical and this is the only manner in which I can circumvent the academic blockade placed on my work and have my work published.

These books I offer for sale on this web page are not yet linguistically edited or controlled and I mention this because in it there is a chance that some grammar errors might lurk in the content. I found the most difficult part of writing is to correct one's own work in grammar because it is done by the method one applies to speak. Please also keep in mind that Afrikaans is my first language and English is my second language. It is also partly in order to have my grammar controlled that I require funding. Part of the funding raised by this effort will go towards editing before I can have them published and it is also for that reason I turn to this effort of marketing the books privately in order to obtain funding to have these books published and marketed through normal channels. I do not intend to challenge William Shakespeare in creating a masterpiece of linguistically prudent magnificence and with Biblical grandeur, but my aim is to get students to see what academics hide from them and to show academics what Newton hides from them. The brainwashing that is going on and the mental control that is inflicted on students is criminal. I challenge any person and every person to prove that I exaggerate or that I show falsified facts.

When any person tells facts that prove to be untrue such a person is a liar notwithstanding the motive in doing it. When any person supplies facts which prove to be untrue with the motive of distorting money in the process, that person are a con artist and an untrustworthy individual. When any person supplies facts which prove to be untrue with the power to black mail students and distort students' mental abilities, such a person is a criminal that should be locked away behind bars.

Again I repeat that academics teach students untruths about Newton that can never be proven and those very same academics know they are teaching blatant lies while being paid to do so. They charge students institution fees only to have students pay for being brainwashed in accepting Newton. Students never dare to challenge Newton's statements because if they do challenge Newton Professors would simply fail their exam papers and banish them from further studying. I dare you to challenge your lecturer just on the facts I give in this web page and see how he or she will react. Academics in physics are paid to minimize your questions, limit student thinking, conceal other ways of reasoning that might be unfavourable to Newton, control all information they give students, have students accept the facts about Newton that they give unconditionally, have students never question or doubt Newton in any manner and accept what they are told to write in examinations.

What do you as a student think would become of the student who asks the professor to prove

$$F = G\frac{M_1 M_2}{r^2}$$

Newton's claims that   is believable when

all evidence points to an expanding Universe?

www.ingramcontent.com/pod-product-compliance
Lightning Source LLC
Chambersburg PA
CBHW080619190526
45169CB00009B/3236